Once Upon a Physical Science Book

12 Interdisciplinary Activities to Create Confident Readers

Jodi Wheeler-Toppen, Karen Kraus, and Matthew Hackett

Arlington, Virginia

Claire Reinburg, Director
Rachel Ledbetter, Managing Editor
Andrea Silen, Associate Editor

Art and Design
Will Thomas Jr., Director

Printing and Production
Catherine Lorrain, Director

National Science Teaching Association
1840 Wilson Blvd., Arlington, VA 22201
www.nsta.org/store
For customer service inquiries, please call 800-277-5300.

Copyright © 2021 by the National Science Teaching Association
All rights reserved. Printed in the United States of America.
24 23 22 21   4 3 2 1

NSTA is committed to publishing material that promotes the best in inquiry-based science education. However, conditions of actual use may vary, and the safety procedures and practices described in this book are intended to serve only as a guide. Additional precautionary measures may be required. NSTA and the authors do not warrant or represent that the procedures and practices in this book meet any safety code or standard of federal, state, or local regulations. NSTA and the authors disclaim any liability for personal injury or damage to property arising out of or relating to the use of this book, including any of the recommendations, instructions, or materials contained therein.

Permissions
Book purchasers may photocopy, print, or e-mail up to five copies of an NSTA book chapter for personal use only; this does not include display or promotional use. Elementary, middle, and high school teachers may reproduce forms, sample documents, and single NSTA book chapters needed for classroom use only. E-book buyers may download files to multiple personal devices but are prohibited from posting the files to third-party servers or websites, or from passing files to non-buyers. For additional permission to photocopy or use material electronically from this NSTA Press book, please contact the Copyright Clearance Center (CCC) (*www.copyright.com*; 978-750-8400). Please access *www.nsta.org/rights-and-permissions* for further information about NSTA's rights and permissions policies.

Library of Congress Cataloging-in-Publication Data

Names: Wheeler-Toppen, Jodi, author. | Kraus, Karen, 1961- author. | Hackett, Matthew, 1978- author.
Title: Once upon a physical science book : 12 interdisciplinary activities to create confident readers / Jodi Wheeler-Toppen, Karen Kraus, and Matthew Hackett.
Description: Arlington, VA : National Science Teaching Association, [2020] | Includes bibliographical references and index. | Identifiers: LCCN 2020025705 (print) | LCCN 2020025706 (ebook) | ISBN 9781681407418 (paperback) | ISBN 9781681407425 (pdf)
Subjects: LCSH: Physical sciences--Study and teaching (Middle school)
Classification: LCC Q183.3.A1 W479 2020 (print) | LCC Q183.3.A1 (ebook) | DDC 500.2071/2--dc23
LC record available at *https://lccn.loc.gov/2020025705*
LC ebook record available at *https://lccn.loc.gov/2020025706*

Contents

Dedication and Acknowledgments vii

About the Authors ix

Chapter 1: Getting Started 1

Chapter 2: The Reading Strategies 11

Chapter 3: The Smash-Masters 23

    Topics: making claims from evidence, designing controlled experiments, using models, car safety testing

    Reading Strategies: comprehension coding, reading in groups

Chapter 4: Identify This 43

    Topics: physical properties, chemical properties, identifying unknowns

    Reading Strategy: chunking

Chapter 5: Handy Heaters 59

    Topics: chemical reactions, chemical equations, chemical hand warmers

    Reading Strategy: reading technical text

Chapter 6: All Charged Up 77

    Topics: static electricity, charges, electric fields

    Reading Strategy: chunking

Chapter 7: Inertia: It's a Drag 91

    Topics: inertia, Newton's first law of motion

    Reading Strategy: previewing diagrams and illustrations

Contents

Chapter 8: Kick Force 105

    Topics: Newton's second law of motion, animal kicks

    Reading Strategy: reading technical text

Chapter 9: Energy's Wild Ride 119

    Topics: energy transformation, conservation of energy, potential and kinetic energy, energy efficiency

    Reading Strategy: talk your way through it

Chapter 10: Taking Your Temperature 135

    Topics: heat, temperature, energy and particle movement

    Reading Strategy: signal words for cause and effect

Chapter 11: How to Not Die in Antarctica 151

    Topics: heat transfer, insulation

    Reading Strategy: signal words for compare and contrast

Chapter 12: Ding-Dong Electromagnets 165

    Topics: electromagnets, magnetic fields, doorbells

    Reading Strategy: previewing diagrams and illustrations

Chapter 13: All About Bat Waves 181

    Topics: introduction to waves, wave vocabulary, bat echolocation

    Reading Strategy: finding the meaning of new words

Chapter 14: Lighting the Way 197

    Topics: waves in communication, digital and analog systems, fiber optics

    Reading Strategy: talk your way through it

Image Credits 211

Index 213

Dedication

For Jon, Natalie, and Zachary
—Jodi Wheeler-Toppen

For Rick
—Karen Kraus

For Carrie
—Matthew Hackett

Acknowledgments

With thanks to a wonderful group of teachers who
field-tested activities from this book:

Amy V. Gilbert
Griffin Middle School, Smyrna, Georgia

Jodie Harnden
Sunridge Middle School, Pendleton, Oregon

Rachel Munshower
Spencerville Middle School, Spencerville, Ohio

Molly Niese
Arlington Local Schools, Arlington, Ohio

Kelli Annin, Jennifer Bahr, Monique Mabin, Gina Oles, and Ben York
Blue Springs School District, Blue Springs, Missouri

And to the following individuals who took time to help us get the science right.
Any mistakes that persist are ours, not theirs!

Elizabeth Gardner
Department of Criminal Justice, University of Alabama at Birmingham

Continued

Acknowledgments

David Hu
School of Mechanical Engineering, Georgia Institute of Technology

Jay Rotella
Weddell Seal Project, Montana State University

Harvey Toppen
Retired Engineer, Formerly with Pratt & Whitney

David Zuby
Vehicle Research, Insurance Institute for Highway Safety

About the Authors

Jodi Wheeler-Toppen is the author of more than 10 science books for children and teachers, including the *Once Upon a Science Book* series with NSTA Press. She has been working in K–12 and teacher education for more than 15 years, with a focus on helping students read and write about science. She loves having adventures with kids—her own and any others who come her way.

Karen Kraus raised and supported a family before teaching middle school science for 13 years in Blue Springs, Missouri. Now in semi-retirement, she continues to work at helping new and experienced teachers improve their instruction. She enjoys traveling to see family, spoiling her grandchildren, and experiencing the wonders of nature.

Matthew Hackett has taught various disciplines for 19 years, including science, mathematics, and reading. He currently teaches sixth- and eighth-grade science in Blue Springs, Missouri. He and his wife coach and volunteer with their church youth group to fill the hours between school days. Matthew is a cat person, but his students remember him for his more exotic classroom pets, including tarantulas, geckos, and snakes.

Chapter 1

Getting Started

In faculty meetings, the principal of my (Jodi's) school would periodically exhort those of us who were "subject-area teachers" to contribute to the school-wide emphasis on improving reading. I would return to my classroom and assign pages from the textbook to my students, only to be greeted by moans and groans. After I finally cajoled my groaners into reading, I would ask questions, but they never seemed to learn much from what they read. Does this sound familiar?

I finally returned to graduate school to find out more about how science teachers could design successful reading lessons for their classes. I compiled what I had learned into a book for life science teachers, which became the first volume in the *Once Upon a Science Book* series. I quickly discovered that I wasn't the only teacher who wished for such a

Getting Started

> **A NOTE ON STANDARDS**
>
> Each chapter in this book is correlated with the *NGSS* and the reading and writing portions of the *CCSS*. At the beginning of each chapter, you will find a detailed chart showing which standards are addressed and what activities within the chapter address those standards. Note that these chapters are not intended to be entire units. Rather, they fit within units and, as such, work well with an *NGSS*-based phenomenon approach to teaching. Depending on your unit, these lessons could be used in many different ways. For example, if your unit is exploring the phenomenon of dropping an aid package from an airplane, Chapter 7 might be used near the beginning to help students understand the role of inertia. If you are looking for a phenomenon related to heat transfer, the engineering activity in Chapter 11 might be expanded to address several heat-related topics.

book—indeed, teachers of other subjects routinely ask me for a book in their area that facilitates reading. With the *Common Core State Standards* (*CCSS*), even more teachers are finding themselves struggling to integrate literacy and science.

The good news is that there are many parallels between how people learn science and how they learn to become better readers. In fact, many studies have indicated that integrating reading and science can lead to gains in both areas (e.g., Fang et al. 2008; Kamil and Bernhardt 2004; Morrow et al. 1997; Radcliffe et al. 2008). Many of the skills you already encourage as a science teacher are important for success in reading, including observing, predicting, making inferences, determining cause and effect, and drawing conclusions (Wallace and Coffey 2019). Thus, you are uniquely positioned to take on the role of coaching your students in content-area reading!

Each lesson in this book consists of a science activity, a reading about an important physical science concept (as determined by the *Next Generation Science Standards* [*NGSS*]), a writing activity that asks students to connect what they did with what they read, and a Thinking Mathematically activity that helps them see how these science concepts connect with mathematics. This book also contains information on teaching specific reading strategies to help you create a complete reading program in your science class.

Learning Science: The Learning Cycle

Science teachers know that people learn science best when they anchor their learning in firsthand experiences. This idea has been formalized into the *learning cycle*, a way to organize lessons so that students have a chance to explore a concept before they learn the relevant vocabulary and principles (Lawson 2009). While the 5E learning cycle is often used in science (see Bybee et al. [2006] in the bottom Find Out More box), in this book we will focus on the critical middle three phases because they relate most directly to integrating reading, and they form the heart of a learning cycle in both science and reading (Baker 2004).

Exploration. In this phase, students experience a new concept in a concrete way by exploring a specific example. Students may be asked to design an experiment, attempt to solve a problem, or simply make observations. Explorations in this book include building an electromagnet, designing an experiment to find the relationship between mass and acceleration, and testing the conductivity of a variety of surfaces. In the exploration, students form an internal framework for the broader ideas that they will encounter during the explanation. This phase also fosters student questions about the topic so they are more interested and engaged during the explanation phase.

Explanation. In this phase, students learn vocabulary and general principles. For example, if students built and made observations of a wave machine during the exploration phase, they would be introduced to the terms *wavelength*, *frequency*, and *amplitude* during the explanation phase. They would also learn that the concept they observed during the exploration—that the wave machine moves energy but not particles—is a general principle that holds for all sorts of waves. Students use that information to build further explanations for what they observed in their exploration. In a learning cycle, explanations can come from lectures, discussions, readings, or videos. For the cycles in this book, the information is provided through reading.

Concept Application. Finally, students need a chance to apply the new terms and principles themselves in a new situation. The concept application phase of a learning cycle can include doing additional hands-on activities, designing new investigations, making a concept map, or solving a new problem. In this book, the concept application phase for each cycle includes a writing prompt and a mathematics activity. In some chapters, it also includes making an argument using a claims, evidence, and reasoning approach.

The learning cycle is based in constructivism, a view of learning that holds that students base new knowledge on the understandings they already hold (Lawson 2009). The prior knowledge that students bring to a lesson may be helpful for learning the new material, or students may bring misconceptions that make learning more difficult. The exploration phase can help challenge students' misconceptions so they are ready to restructure their understanding.

The exploration also fills in gaps that may be present in a student's prior knowledge. For example, if a student's main exposure to heat has been limited to how hot or cold something is, he or she will find it difficult to understand that different materials have different conductive properties. Observing and conducting tests on different materials during the exploration phase can provide the necessary background for learning about the nature of heat transfer.

> **FIND OUT MORE**
> If your students are not familiar with using a claims, evidence, and reasoning approach to argumentation, or if you would like to improve your instruction in that area, see the Argumentation Toolkit at *www.argumentationtoolkit.org*.

> **FIND OUT MORE**
> If learning cycles are new to you, see *The BSCS 5E Instructional Model* (Bybee et al. 2006). This report, prepared for the National Institutes of Health, can be found at *https://media.bscs.org/bscsmw/5es/bscs_5e_full_report.pdf*.

Developing Reading Skills: What Reading Teachers Know

There are several current models of reading comprehension, but they all depict reading as a process of creating a representation of the text in the reader's mind. Many factors play into successful comprehension, including the reader's skill, beliefs about reading, and prior knowledge of the subject of the text (Kendaou, Muis, and Fulton 2011). Reading teachers address these areas using an instructional plan that is very similar to the learning cycle. Reading research conceptualizes a reading lesson as an upside-down pyramid—students need the most support during pre-reading and less support as they move through the other phases (Berkeley and Barber 2015).

Pre-Reading. During pre-reading, reading teachers help students think about what they already know that may be important for understanding the text. For example, if students are going to read about a volcano, the teacher may ask what students know about volcanoes or have them draw a picture of a volcano. Reading teachers call this *activating prior knowledge*. In some cases, reading teachers need to introduce new information that students will need for the reading, such as sharing about the time period in which a story is set or showing students an example of an object that plays a key role in the text.

Reading. This step includes the obvious task of reading the passage, but reading teachers may expand it by asking students to use specific strategies to monitor their comprehension as they read.

Post-Reading. After reading, students are led in reflecting on what they learned and applying this knowledge in a new situation. Reading teachers often have students summarize main ideas, create concept maps, or do projects based on what they read.

A Natural Fit

You can see how the work of science teachers and reading teachers fits well together. An exploration can serve as a pre-reading activity by generating background that supports new learning. It also provides an authentic purpose for reading. No more reading just to answer questions at the end of the chapter—now students can read to answer real questions that they have developed from experiencing science firsthand. For the next phase of the learning cycle, reading provides an excellent source for concept introduction. Reading also models the work of real scientists, who usually read a great deal in the process of developing and interpreting experiments. Finally, both models call for an activity in which students apply their developing knowledge. Figure 1.1 displays the literacy learning cycle used in this book.

One difference in the two models is that reading educators focus more of their attention on what takes place during the actual reading. They know that good readers monitor their comprehension as they read. Reading teachers help students pay attention to whether or not they understand and teach strategies that improve comprehension and memory. Science educators can also teach these strategies, and once again, reading research is on our side—research has shown that students learn reading strategies best if the strategies are incorporated into meaningful reading opportunities (Baker 2004; Fielding and Pearson 1994).

Chapter 1

Figure 1.1. Structures of Learning Cycles and Reading Lessons Fit Together to Form Literacy Learning Cycle

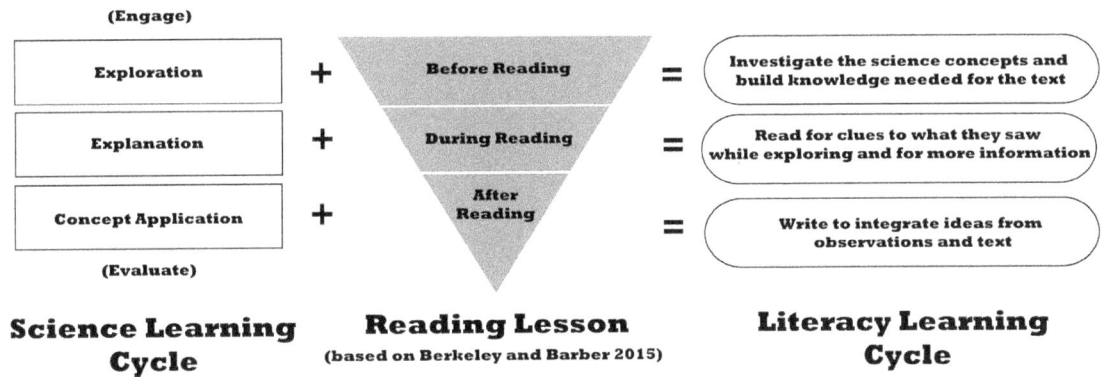

The Larger Goal

It is important to remember when teaching reading in science that our goal is bigger than reading comprehension. We do not want our students to make sense of just one bit of text; we want them to draw from several texts, from their prior experiences, from conversations, from lab, from lecture, and from media and video and to assemble it all into a deep understanding of the topic. We want them to read as an act of "knowledge building" (Cervetti and Hiebert 2015).

Making the Most of This Book

One way to use this book would involve selecting certain lessons that suit your curriculum and using them à la carte. However, this book can also be used as a reading development program by spacing the lessons throughout the year and including the following components that are provided with each chapter:

Strategy Introduction. Each lesson in this book includes a specific reading comprehension strategy that you can introduce before students read. Teaching this strategy does not need to take a great deal of direct instruction; much of the strategy learning will take place during individual practice and small-group interactions. However, explicitly addressing reading strategies can help students learn to take control of their reading (Baker 1991; Radcliffe et al. 2008; Spence, Yore, and Williams 1999).

Reading Groups. Working in reading groups can be a powerful tool for improving comprehension (e.g., Rosenshine and Meister 1994; Schünemann et al. 2017; Wheeler-Toppen 2006). By working together to read a

Getting Started

passage, students can fill in gaps in prior knowledge and model reading strategies for each other (Robb 2000). Having to talk about what they have read causes students to observe how successfully they are comprehending, and over time, that awareness can lead to self-regulation even outside of the group (Baker 2004).

There are a number of ways to organize successful reading groups, and in the next chapter we introduce one such procedure that is simple enough to be implemented in most science classrooms.

Journaling. It takes time, practice, and reflection for a new reading strategy to become a stable part of a student's repertoire. Reading teachers often conduct individual reading conferences with their students to reinforce new strategies, monitor progress, and help students reflect on their development as readers (Allen 1995; Robb 2000; Schoenbach et al. 1999). Science teachers rarely have time to implement individual reading conferences in their classes. Therefore, this book includes journal questions to encourage students to internalize the strategies introduced in class. These questions are designed to help students plan ways to use a strategy, practice a strategy by writing an example situation, or evaluate their own use of a strategy. You can increase the value of the journaling activity by periodically responding to entries in your students' journals.

Assessing Student Learning

Assessment is a critical piece of any teaching endeavor. The assessment exercises in this book are based on the idea that assessments should give you feedback on what your students are learning and also serve as learning opportunities for your students (NRC 1996). Therefore, each activity, in addition to being a learning tool, is designed to provide information about how well your students understand the lesson. These assessments fall into four general categories:

The Big Question. The Big Question is a reading comprehension check found at the end of each reading selection. The question can be answered in a fairly straightforward manner from the information in the text but cannot be answered simply by skimming or quoting a section of text. Answers to the Big Question are generally short, and if students work in reading groups, each group may submit just one answer. You should be able to scan the answers relatively quickly to determine if students grasped the main ideas of the text.

Thinking Mathematically. Physical science offers many opportunities to consider the *NGSS* crosscutting concept of Scale, Proportion, and Quantity (NGSS Lead States 2013). For the most part, the articles in this book tackle topics conceptually, but adding a mathematical component, whether solving problems or simply thinking about a topic in terms of quantities or magnitudes, is an important foundation for future learning of physics and chemistry. Therefore, each chapter has a Thinking Mathematically activity to address these ideas.

Writing Prompt. The writing prompt found in each chapter asks students to integrate what they learned from the exploration and the text. Each activity focuses on a key idea. That idea will be identified as a "key evaluation point" listed below the writing prompt in each chapter. You can use the rubric in Table 1.1, along with this evaluation point, to assess your students' responses to the writing prompt. Because these are open-ended questions, they provide an especially good opportunity to watch for misconceptions that your students may have developed.

Table 1.1. Rubric for Evaluating Responses to Writing Prompts

	Completely	Partially	Incorrectly or Insufficiently
Does the response correctly describe the key idea?			
Is the response supported by details from the reading and/or investigation?			
What misconceptions are present in the response?			

Claims, Evidence, and Reasoning. Many of the activities in this book ask students to make a claim and support it with evidence. The rubric in Table 1.2 (p. 8) can be used to assess the claims that students make in these activities and also serve as a springboard for a class discussion about what makes an effective claim. As students grasp the basic concepts, you might consider moving to a more sophisticated tool to help them think through their claims, such as the Argumentation and Evaluation Guide, developed by Bulgren and Ellis (2015).

Additionally, you will want to assess how students are developing as readers. You will find information on assessing reading skills at the end of Chapter 2.

Remember that assessment should affect what you do next with your students (Black 2003). If you find that students are struggling with the Big Question or Thinking Mathematically sections, you may need to talk about the text with your class. If you encounter misconceptions in responses to the writing prompts, you need to address them. Conversely, if students are successful with these activities, you can move on with the confidence that they are ready for new topics.

Getting Started

Table 1.2. Simple Rubric for Assessing Claims, Evidence, and Reasoning

	Completely	Partially	Not at All
Claim			
• Is it the type of claim that an experiment could verify?			
• Does the claim address the question asked in the investigation?			
Evidence			
• Does the evidence support the claim?			
• Is the evidence sufficient to support the claim—that is, do you have enough information?			
• Is the claim correct based on the data from the investigation?			
Reasoning			
• Does the explanation adequately connect the claim and the evidence?			

IMPORTANT SAFETY INFORMATION

With hands-on, process, and inquiry-based activities and investigations, the teaching and learning of science today can be both effective and exciting. However, potential safety issues must be addressed by conducting hazard assessments and by using appropriate engineering controls (e.g., ventilation, fume hoods, fire extinguishers, showers), administrative and safety operating procedures, and appropriate personal protective equipment (e.g., indirectly vented chemicals-splash goggles that meet ANSI/ISEA Z87.1 D standard; chemical-resistant, nonlatex aprons; nonlatex gloves).

Teachers can make it safer for students and themselves by adopting, implementing, and enforcing legal safety standards and better professional safety practices in the science classroom and laboratory. Throughout this book, safety notes are provided for investigations and activities, and they need to be adopted and enforced in an effort to provide for a safer learning and teaching experience. Teachers should also review and follow local polices and protocols used within their district and/or school (e.g., a chemical hygiene plan, board of education safety policies).

The National Science Teaching Association provides additional applicable standard operating procedures here: *http://static.nsta.org/pdfs/SafetyInTheScienceClassroom.pdf.* Furthermore, an acknowledgement form for middle school students can be found at *http://static.nsta.org/pdfs/SafetyAcknowledgmentForm-MiddleSchool.pdf.* Students should be required to review the document or one similar to it under the direction of the teacher. Both students and their parent/guardian should then sign the document acknowledging procedures that must be followed for a safer working and learning experience.

Disclaimer: The safety precautions of each activity are based in part on the use of the recommended materials and instructions, legal safety standards, and better professional practices. Selection of alternative materials or procedures for these activities may jeopardize the level of safety and therefore is at the user's own risk.

Get Started!

With this book, you have everything you need to boost your students' science and reading skills. Learn about the strategies you need for the book in Chapter 2. Then dive into the 12 content chapters. As you and your students work through these lessons together, you will be able to watch their confidence as readers—and your confidence as a reading educator—grow. So what are you waiting for? Let's get started!

References

Allen, J. 1995. *It's never too late: Leading adolescents to lifelong literacy.* Portsmouth, NH: Heinemann.

Baker, L. 1991. Metacognition, reading, and science education. In *Science learning: Processes and applications*, ed. C. M. Santa and D. E. Alvermann, 2–13. Newark, DE: International Reading Association.

Baker, L. 2004. Reading comprehension and science inquiry: Metacognitive connections. In *Crossing borders in literacy and science instruction*, ed. E. W. Saul, 239–257. Newark, DE: International Reading Association.

Berkeley, S., and A. T. Barber. 2015. *Maximizing effectiveness of reading comprehension instruction in diverse classrooms*. Baltimore, MD: Brookes Publishing.

Black, P. 2003. The importance of everyday assessment. In *Everyday assessment in the science classroom*, ed. J. M. Atkin and J. E. Coffey, 1–11. Arlington, VA: NSTA Press.

Bulgren, J., and J. Ellis. 2015. The argumentation and evaluation guide: Encouraging NGSS-based critical thinking. *Science Scope* 38 (7): 78–85.

Bybee, R. W., J. A. Taylor, A. Gardner, P. Van Scotter, J. C. Powell, A. Westbrook, and N. Landes. 2006. *The BSCS 5E instructional model: Origins, effectiveness, and applications*. Colorado Springs, CO: BSCS. https://media.bscs.org/bscsmw/5es/bscs_5e_full_report.pdf.

Cervetti, G. N., and E. H. Hiebert. 2015. Knowledge, literacy, and the *Common Core*. *Language Arts* 92 (4): 256–269.

Fang, Z., L. Lamme, R. Pringle, J. Patrick, J. Sanders, C. Zmach, S. Charbonnet, and M. Henkel. 2008. Integrating reading into middle school science: What we did, found, and learned. *International Journal of Science Education* 30 (15): 2067–2089.

Fielding, L. G., and P. D. Pearson. 1994. Reading comprehension: What works. *Educational Leadership* 51 (5): 62–68.

Kamil, M. L., and E. Bernhardt. 2004. The science of reading and the reading of science: Successes, failures, and promises in the search for prerequistie reading skills for science. In *Crossing borders in literacy and science instruction*, ed. E. W. Saul, 123–139. Newark, DE: International Reading Association.

Kendeou, P., K. R. Muis, and S. Fulton. 2011. Reader and text factors in reading comprehension processes. *Journal of Research in Reading* 34 (4): 365–383.

Lawson, A. 2009. *Teaching inquiry science in middle and secondary schools*. Thousand Oaks, CA: SAGE Publications.

Morrow, L. M., M. Pressley, J. K. Smith, and M. Smith. 1997. The effect of a literature-based program integrated into literacy and science instruction with children from diverse backgrounds. *Reading Research Quarterly* 32 (1): 54–76.

National Research Council (NRC). 1996. *National science education standards*. Washington, DC: National Academies Press.

NGSS Lead States. 2013. *Next Generation Science Standards: For states, by states*. Washington, DC: National Academies Press.

Radcliffe, R., D. Caverly, J. Hand, and D. Franke. 2008. Improving reading in a middle school science classroom. *Journal of Adolescent and Adult Literacy* 51 (5): 398–408.

Robb, L. 2000. *Teaching reading in middle school: A strategic approach to reading that improves comprehension and thinking*. New York: Scholastic Professional Books.

Rosenshine, B., and C. Meister. 1994. Reciprocal teaching: A review of the research. *Review of Educational Research* 64 (4): 479–530.

Schoenbach, R., C. Greenleaf, C. Cziko, and L. Hurwitz. 1999. *Reading for understanding: A guide to improving reading in middle and high school classrooms*. New York: Jossey-Bass.

Schünemann, N., N. Spörer, V. Völlinger, and A. Brunstein. 2017. Peer feedback mediates the impact of self-regulation procedures on strategy use and reading comprehension in reciprocal teaching groups. *Instructional Science* 45 (4): 395–415.

Getting Started

Spence, D. J., L. D. Yore, and R. L. Williams. 1999. The effects of explicit science reading instruction on selected grade 7 students' metacognition and comprehension of specific science text. *Journal of Elementary Science Education* 11 (2): 15–30.

Wallace, C., and D. Coffey. 2019. Investigating elementary preservice teachers' designs for integrated science/literacy instruction highlighting similar cognitive processes. *Journal of Science Teacher Education* 30 (5): 507–527.

Wheeler-Toppen, J. 2006. Reading as investigation: Using reading to support and extend inquiry in science classrooms. PhD diss., University of Georgia, Athens.

Chapter 2
The Reading Strategies

What Is Reading?

At lunch, my colleagues and I (Jodi) would periodically bemoan how our students "can't read." But what did we really mean by that? Certainly, most of our students, even those scoring well below grade level on reading tests, could pronounce the words on the page of a simple book. Some of them even enjoyed reading novels for fun. But they seemed completely unable to make sense of their science textbooks or other school books.

Part of the problem lay in the way our students thought about reading. Many believed that reading consisted of

The Reading Strategies

calling out the words on the page. Good readers, they assumed, automatically understood all those words. For struggling readers, their failure to understand reinforced their beliefs that they didn't read well. Reading strategies can be important for helping students improve their reading, but students need something more. They need to begin to view reading as an active search for meaning that is *within their control*. We can change how our students think about reading through the way we talk about reading in our classrooms.

Starting the Conversation

The first step is to create a classroom culture in which students feel safe exploring new ways of thinking about reading. You can begin by simply stating aloud that reading can be difficult, even for good readers. You can share your own stories about encountering words, phrases, or books that were hard for you to understand. Most important, you should make it clear that you will not tolerate students teasing each other about reading struggles.

Next, students need a chance to see what experienced readers do as they read. Good readers constantly monitor their comprehension and notice if they do not understand what they read. They have an ongoing conversation in their head in which they compare what they are reading with what they already know (and sometimes argue with the text if they disagree). When they do not understand, or they find inconsistencies with their prior knowledge, they use problemsolving strategies to make sense of the text or resolve the inconsistencies. All these things are hidden from someone watching, but as teachers we can make them visible.

Think-Alouds. One way to make visible the invisible processes of reading is to talk about what we are thinking as we read (Baker 2004; Kucan and Beck 1997). This is called a think-aloud. For example, you might read this section from the article in Chapter 9:

> *Remember that energy is the ability of an object to do work. When an object such as a roller coaster train is moving, it's easy to picture how it could do work by, say, smashing into an empty cart blocking the track. The energy of an object in motion is called kinetic energy. But what about a train at the top of the first roller coaster hill, sitting still for just a moment before plunging down? What kind of energy could it have?*

To use this in a think-aloud, you would insert your own thoughts as you read out loud. You can introduce the think-aloud by saying to your students, "Have you ever noticed that when you read, there's one voice in your head saying the words on the page and another voice talking about what you are reading? I'm going to read a passage and try to tell you what both of my voices are saying." Then your think-aloud might sound something like this:

> *Remember that energy is the ability of an object to do work.* (Let's see, we've talked about the meaning of the word *work* in science. It's when a force causes something to move.) *When an object such as a roller coaster train is moving, it's easy to picture how it could do work* (Actually, I'm not sure I can picture that.) *by, say, smashing into an empty cart blocking the track.* (Oh, that's a good example. I can see that.) *The energy of an object in motion is called* kinetic energy. (I don't completely get the phrase kinetic energy, but I'm betting it's something I need to keep watching for as I read.) *But what about a train at the top of the first roller coaster hill, sitting still for just a moment before plunging down? What kind of energy could it have?* (Wow, that doesn't seem like it would have any energy at all. But this is a question, so I'm guessing the next sentence will answer the question. I'll look for the answer as I read.)

This allows struggling readers to "see" how strong readers approach difficult reading passages. You can use a think-aloud to demonstrate specific strategies to your class or when you are helping a student or small group figure out a confusing passage.

Peer Conversation. Students can also show invisible aspects of reading to each other. In reading groups (see discussion on p. 14), students share with each other how they made sense of the text. One student might read the above passage and ask the group, "What does it mean that energy can change form?" Another student might add, "And it says energy can't be used up, but a roller coaster stops eventually." A third student might say, "Maybe the next few sentences will give us more information." You may be skeptical that your students would be able to have these discussions with each other. You will be surprised. As students get used to working in groups, and as an atmosphere of trust develops, even weak readers become comfortable asking for help and sharing what they think as they read.

Overarching Strategies

Each lesson in this book can be used to introduce students to one or more specific reading strategies. These strategies are important; they represent ways that good readers solve specific reading problems. However, remember that learning specific strategies is not the ultimate goal. We want students to begin to approach reading as an active search for meaning (Cervetti and Hiebert 2015; Loxterman, Beck, and McKeown 1994). The following strategies are only a means to that end.

The Reading Strategies

The first two strategies introduced here, comprehension coding and reading groups, are intended to be used throughout the lessons. They address the primary issues of comprehension monitoring and problem solving.

Comprehension Coding. In comprehension coding, students mark codes to indicate what they are thinking as they read. We recommend introducing the following codes:

! This is important.
✓ I knew that.
X This is different from what I thought.
? I don't understand.

Over time, students may develop their own coding systems that meet their particular needs. Indeed, you may notice that you do something similar yourself. You may underline important information you want to remember or jot questions in the margins of books. This strategy is intended to mimic that sort of behavior on the part of good readers and encourage students to monitor their comprehension as they read.

Reading Groups. Working in reading groups can be a powerful tool for improving comprehension (e.g., Rosenshine and Meister 1994; Wheeler-Toppen 2006). There are a number of ways to organize reading groups; however, we recommend the following simple procedure for the activities in this book.

In this procedure, each reading group has three students, and each member will lead the group through one section of an article using the procedure in Figure 2.1. Each student will need a card telling him or her what section of the reading to lead and the procedure for leading. Furthermore, each group will need an orange flag—either a folded piece of

Figure 2.1. Reading Group Procedure

You are the leader for Section _____.
What to do in your group:
- Everyone should **read the section and use codes** (✓, X, ?, and !) to mark it.
- The leader for this round **tells what the section was about**. If you're stuck, try starting with, "What I understand so far is …"
- **Ask if anyone found something confusing** (marked X or ?).
  ◊ The group should work together to figure out what the confusing word, sentence, or idea means.
  ◊ If the group cannot make sense of it, raise your orange flag for help.
- **Turn to the next leader** and **repeat for the next section**.
- When your group has read and discussed all three sections, work together to **answer the Big Question**.

construction paper that is propped up or a plastic cup—to place on the desk when help is needed.

This group procedure calls for readings to be broken into three sections. We've divided all the articles in this book into different segments using short black lines. That way, articles can be used with the group-reading procedure whenever you see fit.

Your role as the teacher is important during this process. Initially, you will need to monitor students closely to ensure that they really do follow the procedure. When students raise their flags, listen to their comprehension difficulty. What have they tried so far? Can you model a strategy for making sense of the text? Is there a piece of background knowledge or a word meaning that you need to provide?

Listen in on groups that are not having trouble as well. Encourage students to share their strategies. For example, if one student tells another what a sentence means, ask, "How did you know that?" By participating in the groups with students, you show that even teachers must think carefully about what they read.

As with any classroom procedure, this one takes practice. For the first few sessions, students will have to focus as much on what to do as on what they are reading. They will also need to see that you are serious about requiring them to follow the procedure. After two or three sessions, however, students will begin to follow the procedure automatically and be able to focus more on content. The time invested in learning the process will be well worth it, as students' reading skills and confidence improve. Step-by-step instructions for introducing reading groups can be found in Chapter 3.

> **TEACHING NOTE**
> Some teachers have difficulty using comprehension coding with their students because their school restricts the number of copies they can make. If you are in this position, consider the following ideas:
> - Talk to your administrator about the problem. Most administrators are interested in supporting attempts to improve reading. They may be willing to allow you extra copies for this purpose.
> - Consider printing a class set of readings and having them laminated or slipped into clear page protectors. Students can code using overhead projector pens and then wipe off their marks for the next class. These class sets can be used for several years.
> - To use this strategy with textbooks, you can give students strips cut from sticky notes to mark sentences as they read.

Problem-Solving Strategies

The rest of the strategies in this book are designed to help students solve specific comprehension problems or learn something about how science texts are organized. Each chapter will describe how to introduce one of the following strategies that would be appropriate to use while reading the article for that chapter. Before giving the article to your class, read it yourself and identify places the strategy would be useful. This will help you guide your students' reading.

Keep in mind that students need practice to master any strategy. For this reason, it will be important for you to monitor students closely the first time they use a strategy. In this book, several strategies appear in two different lessons to reinforce their use. You may also want to follow up with readings from your textbook or other sources to allow students to practice the strategies further.

The Reading Strategies

> **TEACHING NOTE**
> Sending students to the dictionary, or even the glossary in the back of a book, is not a particularly productive way to help them with unknown words. Dictionary definitions can be more confusing than the original text and include additional words that students do not know. Furthermore, the task of looking up the word interrupts a student's reading process. It is much less disruptive to simply tell students the meaning of a word if it cannot be figured out from the context.

Finding the Meaning of New Words. Specialized vocabulary is a key feature of science texts (Fang 2006; Holliday 1991). As students read, they are continually introduced to new terms. Struggling readers often miss the definitions of the words when they are introduced because they don't recognize the cues that a definition is being given.

Most students will have been introduced to the strategy of using context clues to find the meaning of new words. As a general reading strategy, using context clues means looking at the surrounding text to figure out a likely meaning of the unfamiliar word. Although many students know they should "use context clues," they often do not use the strategy successfully. Chapter 13 helps students practice this skill by focusing on some of the most common ways that new definitions are presented in science text (see Table 2.1). Students can learn to look for the clues in the sentence before and after a new word is used for the first time.

Note that sometimes the text does not provide sufficient context clues for students to figure out the meaning of a word, especially for non-science words or vocabulary that is not the focus of the reading. These words constitute background knowledge that the writer expects students to have already. One advantage to having students in reading groups is that they can help each other with these words. Alternatively, if you can identify words that may cause problems, you can teach the words before giving students the text.

Table 2.1. Common Ways That Texts Introduce New Words

Example	Explanation
A bat's vocal cords produce vibrations. Vibrations are quick shaking motions that move back and forth.	The sentence after the term provides a definition.
A bat's vocal cords make quick shaking motions that move back and forth, called vibrations.	The new term is signaled with the word *called*.
A bat's vocal cords produce vibrations, which are quick shaking motions …	The definition is signaled with the phrase *which are* or *which means*.
A bat's vocal cords produce vibrations, or quick shaking motions that move back and forth.	The word *or* after a comma indicates that the vocabulary word and the term or phrase that follows *or* mean the same thing. (This is a signal that many students miss!)
A bat's vocal cords produce vibrations. These quick shaking motions …	This is the trickiest situation. The text doesn't directly say what the word means but implies it by using the word and definition close together.

Previewing Diagrams and Illustrations. Learning from text that includes images requires integrating the information from both sources, and readers who spend time using both sources of information learn more successfully from the text (Butcher 2006; Mason, Tornatora, and Pluchino 2013). However, many students ignore diagrams and illustrations (Wheeler-Toppen 2006). By doing this, they miss out on excellent reading support and, often, on key information (Holliday 1991; Vasquez, Comer, and Troutman 2010).

Diagrams and illustrations can clarify points that are hard to explain in words. They may provide useful background knowledge or give examples. Pictures can be especially helpful for struggling readers because they provide hints about what the text will say. Students who come to the text with weaker prior knowledge spend even less time studying diagrams and have more difficulty integrating the text with the images (Ho et al. 2014). Therefore, many students will need support to tackle diagrams in their science reading.

Previewing diagrams and illustrations can serve several functions. When you, as the teacher, lead the preview, you can use it to call attention to important points that the text will make. You can also use it to help students draw on their prior knowledge by having them look for things that they recognize in the pictures. Previewing can also be a useful tool for helping students make predictions before they read. When students begin reading, they will attempt to find out if their predictions were correct. Chapters 7 and 12 take advantage of these possibilities.

Text Signals. Some texts expect readers to infer connections between ideas as they read (e.g., "The cat ran inside. It started to rain."). Other texts do a better job of explicitly identifying those connections (e.g., "The cat ran inside because it started to rain.") We know that it is much easier to understand text that explicitly makes the connections between ideas (Graesser, McNamara, and Kulikowich 2011; Hall et al. 2016). It stands to reason, then, that students would do well to pay attention to the signal words that indicate connections in the text.

Some teachers have compared these text signals to traffic signs (Schoenbach et al. 1999). When readers see these words or phrases, they should slow down and notice the information that follows. Chapters 10 and 11 focus on two common groups of signal words. Chapter 10 looks at signals for cause and effect, and Chapter 11 presents signals for comparing and contrasting (see Table 2.2, p. 18).

Although these are the only signal words that are introduced formally in this book, there are many other types. For example, words such as *following*, *previously*, and *during* indicate that a timeline is being given. The presence of a question in a text generally indicates that an answer will be given. Signals such as *like, including, such as, for instance*, and *e.g.* tell the reader that examples are being given. You may find opportunities to introduce these or other text signals in your conversations with students about reading.

Table 2.2. Text Signals Introduced in Chapters 10 and 11

Cause and Effect	*because, cause/are caused by, consequently, so, as a result, therefore, for this reason, thus, hence, in response to,* and *leading to* (*since* often indicates cause and effect)
Comparisons	*similarly, in the same way, just like, just as, likewise, too,* and *also*
Contrasts	*however, in contrast, on the other hand, conversely, whereas, alternatively,* and *instead* (*but, yet,* and *while* sometimes indicate a contrast)

Chunking. Science texts often have a lot of information crammed into each sentence (Fang 2006). Consider the following sentence about waves:

The wavelength of a longitudinal wave, such as a sound wave, is the distance between the compressions.

To understand the subject of this sentence—the wavelength of longitudinal waves—the reader has to process that the sentence is about waves in the scientific sense, think about what a wavelength is, and remember that longitudinal waves are the ones that move in the same plane as particle movement. And that's just to get through the subject, before the reader even thinks about the main point of the sentence!

The example sentence also includes the phrase *such as a sound wave*, which is stuck in the middle of the main thought. This type of phrase, which is common in science writing, is called the interruption construction (Fang 2006) because it interrupts the flow of the sentence. If the interruption is long, it can be especially confusing for struggling readers.

Experienced readers intuitively break sentences like the one above into chunks and think about each chunk individually. Struggling readers may try to understand the whole sentence at one time (Schoenbach et al. 1999). The reading strategy of chunking helps students break sentences into separate ideas. Chapters 4 and 6 introduce students to chunking by showing them that they can stop to think as they read, even when there is no period or comma to signal a pause.

Talk Your Way Through It. Sometimes students will tell us that they read their book but can't remember what they read. They may be working so hard to understand individual sentences that they fail to grasp what the text as a whole means. Similarly, the text may contain so much information that it is difficult to remember. Talk your way through it is essentially a summarizing strategy, but we find that students are less intimidated by the phrase *talk your way through it* than they are by the idea of doing something as formal sounding as summarizing.

In the talk your way through it strategy, students stop throughout the text to rehearse the information they have just read. They may need to use a sentence starter such as "This section is saying that …" or "What I understand so far is …" If they have difficulty stating it in their own words, then they know they need to reread the section for clarification. If you use the group reading procedure with your students, they will be doing a version of talk your way through it with each section, but Chapters 9 and 14 address the strategy directly and give suggestions for talking about it with your students.

Reading Technical Text. The final set of strategies in this book deals with the special challenge of reading text with chemical and mathematical equations. Symbols and equations add an extra level of challenge to reading science. In fact, even students who are successful at reading other formats can struggle with texts that embed mathematical symbols and equations (Österholm 2006).

Text that includes equations assumes that the "implied" reader will do certain things as they read, including interpreting the meaning of symbols, solving problems as they read, and generalizing important ideas from the examples given. Most real-life readers don't demonstrate these skills (Weinberg and Wiesner 2011). If students are going to be successful with technical texts, we must help them learn to read them.

In this book, we use the phrase *back-and-forth reading* to talk to students about reading text with chemical or mathematical equations. The idea is that students should compare the written explanation to the symbols as they read, saying the word meanings of the symbols to themselves as they go. Additionally, if there are sample problems, they should work through the problems, making sure they understand how the problem progresses from line to line. Details on introducing students to reading technical texts can be found in Chapters 5 (chemical reactions) and 8 (mathematics).

Using Self-Assessment to Monitor Strategy Development

Reading teachers often use individual and small-group conferences to assess how students are developing as readers. Such conferences are beyond the scope of what most science teachers are able to incorporate into their classes. Therefore, you can monitor your students' strategy use and reading confidence by using a periodic self-assessment. To conduct the assessment, have students answer the following questions in their reading journals (adapted from Robb 2000):

The Reading Strategies

- What do you do before reading to get ready to learn?
- While reading, what do you do if you come to a word or section that you do not understand?
- How do you help yourself remember the details of your reading?
- What would you like to do better as a reader?

Then give students a slip of paper with the chart in Table 2.3 to complete and tape into their journals. Give the assessment first at the beginning of the year, before you start to teach the strategies. Read over your students' answers, as you may want to make changes to your instruction based on what they say. Give the assessment again midyear and at the end of the year, and encourage students to look over their previous responses and note how they have improved.

Table 2.3. Self-Assessment of Strategy Use

How Often Do I ... ?	A Lot	Sometimes	Never
Use codes (such as ✓, +, ?, and X) to mark what I'm thinking as I read			
Use the information around a new word to figure out what it means			
Study the diagrams and illustrations before reading			
Use text signals to recognize causes and effects			
Use text signals to recognize comparisons and contrasts			
Chunk difficult sentences into smaller pieces to read			
Talk my way through texts that have a lot of new information			
Translate chemical and mathematical symbols into words and solve sample problems as I read			

References

Baker, L. 2004. Reading comprehension and science inquiry: Metacognitive connections. In *Crossing borders in literacy and science instruction*, ed. E. W. Saul, 239–257. Newark, DE: International Reading Association.

Butcher, K. R., 2006. Learning from text with diagrams: Promoting mental model development and inference generation. *Journal of Educational Psychology* 98 (1): 182–197.

Cervetti, G. N., and E. H. Hiebert. 2015. Knowledge, literacy, and the *Common Core*. *Language Arts* 92 (4): 256–269.

Fang, Z. 2006. The language demands of science reading in middle school. *International Journal of Science Education* 28 (5): 491–520.

Graesser, A. C., D. S. McNamara, and J. M. Kulikowich. 2011. Coh-Metrix: Providing multilevel analyses of text characteristics. *Educational Researcher* 40 (5): 223–234.

Hall, S. S., J. Maltby, R. Filk, and K. B. Paterson. 2016. Key skills for science learning: The importance of text cohesion and reading ability. *Educational Psychology* 36 (2): 191–215.

Ho, H. N. J., M. J. Tsai, C. Y. Wang, and C. C. Tsai. 2014. Prior knowledge and online inquiry-based science reading: Evidence from eye tracking. *International Journal of Science and Math Education* 12: 525–554.

Holliday, W. G. 1991. Helping students learn effectively from science text. In *Science learning: Processes and applications*, ed. C. M. Santa and D. E. Alvermann, 38–47. Newark, DE: International Reading Association.

Kucan, L., and I. L. Beck. 1997. Thinking aloud and reading comprehension research: Inquiry, instruction, and social interaction. *Review of Educational Research* 67 (3): 271–299.

Loxterman, J. A., I. L. Beck, and M. G. McKeown. 1994. The effects of thinking aloud during reading on students' comprehension of more or less coherent text. *Reading Research Quarterly* 29 (4): 352–367.

Mason, L., M. C. Tornatora, and P. Pluchino. 2013. Do fourth graders integrate text and picture in processing and learning from an illustrated science text? Evidence from eye-movement patterns. *Computers and Education* 60: 95–109.

Österholm, M. 2006. Characterizing reading comprehension of mathematical texts. *Educational Studies in Mathematics* 63 (3): 325–346.

Robb, L. 2000. *Teaching reading in middle school: A strategic approach to reading that improves comprehension and thinking*. New York: Scholastic Professional Books.

Rosenshine, B., and C. Meister. 1994. Reciprocal teaching: A review of the research. *Review of Educational Research* 64 (4): 479–530.

Schoenbach, R., C. Greenleaf, C. Cziko, and L. Hurwitz. 1999. *Reading for understanding: A guide to improving reading in middle and high school classrooms*. New York: Jossey-Bass.

Vasquez, J. A., M. W. Comer, and F. Troutman. 2010. *Developing visual literacy in science K–8*. Arlington, VA: NSTA Press.

Weinberg, A., and E. Wiesner. 2011. Understanding mathematics textbooks through reader-oriented theory. *Educational Studies in Mathematics* 76 (1): 49–63.

Wheeler-Toppen, J. 2006. Reading as investigation: Using reading to support and extend inquiry in science classrooms. PhD diss., University of Georgia, Athens.

Chapter 3
The Smash-Masters

Topics
- Making claims from evidence
- Designing controlled experiments
- Using models
- Car safety testing

Reading Strategies
- Comprehension coding
- Reading in groups

The Smash-Masters

Connections to Standards

Next Generation Science Standards (NGSS) Correlations		
Dimension	**Element**	**Matching Student Task or Question From the Activity**
Science and Engineering Practice(s)	• Planning and Carrying Out Investigations • Engaging in Argument From Evidence	• Students plan and conduct an investigation measuring the speed of toy cars. • In the writing prompt, students form an argument to support their claim about the speed of toy cars using evidence and reasoning (by communicating the ways they controlled the variables in the experiment).
Crosscutting Concept(s)	• Systems and System Models	• Students read about how scientists use system models for crash test safety.
Common Core State Standards (CCSS) Correlations		
Reading Standard(s)	• CCSS.ELA-Literacy.RST.6-8.1. Cite specific textual evidence to support analysis of science and technical texts. • CCSS.ELA-Literacy.RST.6-8.2. Determine the central ideas or conclusions of a text; provide an accurate summary of the text distinct from prior knowledge or opinions.	• Students are introduced to coding and reading in groups. Coding requires close consideration of textual evidence. The reading-group procedure includes summarizing sections of text.
Writing Standard(s)	• CCSS.ELA-Literacy.WHST.6-8.1. Write arguments focused on discipline-specific content.	• The writing prompt asks students to make a claim using the data from their controlled experiment and support it with evidence and reasoning.

Note: This chapter will help you introduce two skills: comprehension coding and the group-reading procedure. The first time a class uses the group-reading procedure, students' working memory is "tied up" in learning the procedure; they do not learn the text content as well as they will in future group-reading experiences. For this reason, we do not cover a specific performance expectation or disciplinary core idea in Chapter 3. Instead, we focus on a science topic (controlled experiments) that students have probably encountered before.

Background

This chapter has two main goals. The first is to ease students into the overarching strategies described in Chapter 2. For that reason, this chapter has two reading passages. "Fair on the Field—and in the Lab" (p. 29) allows students to practice using codes to record what they are thinking as they read. "The Smash-Masters" (p. 35) can be used to let students practice the group-reading process. If you choose not to have your students work in reading groups, using two reading passages with this lesson will reinforce that reading is an important part of your science class.

The second goal is to review for students how to design a controlled experiment. Scientists working in physics and chemistry use various types of research to answer questions. They may observe natural systems in action and look for patterns. They may program computer models to try to predict outcomes in a variety of situations. They may take a series of measurements to try to establish the value of a constant.

Controlled experiments, however, are a cornerstone of physical science research. Although many students will have had exposure to controlled experiments in prior classes, most will have not mastered it by middle school. In the exploration for this chapter, students will practice designing their own controlled experiments using toy race cars (or other materials that you might have on hand).

Pre-Reading/Exploration
Materials for Activities (Per Group)
- Safety glasses with side shields or safety goggles for each student
- 2 toy cars
- 2 ramps
- 4–6 books or blocks to prop up ramps
- 1 stopwatch or clock with second hand

> **SAFETY NOTES**
> The following safety recommendations apply to all activities in this chapter:
> - Wear safety glasses with side shields or safety goggles during the setup, hands-on, and take-down segments of the activities.
> - Immediately pick up any items dropped on the floor, as they are a slip-and-fall hazard.
> - Appropriately dispose of lab materials at the end of the activities as directed by the teacher.
> - Immediately report any lab accident to the teacher.
> - Only wear closed-toed shoes during these activities to prevent foot injuries.
> - Wash your hands with soap and water after completing the activities.

Activity 1

For this activity, you will need cars, ramps, and blocks or books to prop up the ramps. You may also need masking tape to secure the ramps. For the cars, you could use Matchbox-style cars, creations from plastic construction blocks and wheels, physics carts from a science supply store, or whatever you can easily access. The ramps can be toy tracks, rain gutters, books, blocks of wood, metersticks, or whatever else will fit your cars.

Begin by having students pick out a car that they like and pair up with a partner. Tell each pair to figure out whose car is fastest. Provide them access to ramps and blocks to set up their race, but don't give any further instructions. Allow about five minutes for them to race their cars.

Call the groups back for discussion. Ask the winners to raise their hands, and then follow up with some of the partners: Do you agree that your partner's car was faster? Was your race fair? How do you know? If you did the same race again, do you think your partner's car would win again? Have the class list things they needed to do to make sure their race was fair. This might include having the ramps at the same angle, starting the cars at the same time, not allowing one person to push the car, and making sure both cars were pointed straight forward.

Reading 1

Use With Student Page(s): "Fair on the Field—and in the Lab" (article)

Introduce the First Reading. Tell students that you will come back to racing cars after they have read about keeping things fair in sports and science. Introduce the idea of using codes as described in the Reading Strategies section that follows. Then have students read "Fair on the Field—and in the Lab."

Reading Strategy 1: Comprehension Coding

To introduce the strategy, display the first three paragraphs of "Fair on the Field—and in the Lab" (p. 29) so students can watch as you code. Read the paragraphs out loud and model the coding process. For example, you might read the first paragraph and put a question mark next to it, saying, "This is confusing to me because I don't see how it has anything to do with science." After the description of the track field, you could put a check for "I knew that" and comment that you or someone you know ran track in high school. After the statement about swimming pool lanes, you could put an X and say you thought those lanes were empty so swimmers wouldn't bump against the wall. And after the statement that a controlled experiment is like a sporting event, you could add an exclamation point and say that this seems like it might be one of the big ideas of the article. Point out that you did not put a code next to every sentence; students only need to place codes where they seem appropriate.

After modeling the strategy for a few paragraphs, give students time to read and code the rest of the article independently. After reading, have students share some of their coded sentences and explain why they chose those codes. Make sure any student with an X or question mark has a chance to have their questions or confusions resolved.

Activity 2

Use With Student Page(s): Racing for Control: Designing a Controlled Experiment (lab sheet)

Tell students they are going to return to their toy cars to answer the question, "Which car is faster?" This time, however, they are going to plan a controlled lab experiment. Distribute the lab sheet Racing for Control: Designing a Controlled Experiment (p. 31) and have students use the first few sections to decide on the details of their race. There are a variety of ways students may organize their races. They may construct identical tracks and run several races to see which car reaches the bottom first. They may choose to time each car separately on the same track and compare times. Or they may come up with a different design entirely. Once students have a plan, they can conduct their tests.

The writing assignment for this lesson is at the bottom of the lab sheet in the sections covering claims, evidence, and reasoning. You may wish to have students stop after they collect data and discuss their writing plans (see the Application/Post-Reading/Writing section of this chapter on p. 28).

Reading 2

Use With Student Page(s): Reading-Group Roles (student handout) and "The Smash-Masters" (article)

Introduce the Second Reading. Tell students they are going to read a passage that talks about how controlled experiments are used to make cars safer. You may wish to preface the reading with a video clip of a car crash test. Several are online, including at the Insurance Institute for Highway Safety (IIHS) website (*https://classroom.iihs.org*).

Reading Strategy 2: Reading in Groups

Before you introduce this strategy, prepare cards for each reading group using the Reading-Group Roles student handout (p. 33). It is helpful to use these cards every time you do group reading. Then, if a group has gotten off task, you can ask what section they are on, check their cards, and get the correct leader back on track.

Read over the procedure with your class. Then select three confident students to come to the front of the class and model the procedure for the first section of text in "The Smash-Masters" article (down to the first short black line). Next, place the rest of your students into groups of three to try the process themselves. Have them start from the beginning of the article to reinforce the process they just watched.

Remember that the first time students use a procedure like this, it may not go smoothly. You will need to monitor them closely and insist that students participate and follow the process. Likewise, the first time the procedure is used, students are concentrating as much on figuring out what to do as they are on reading. Students will benefit most from the process after they have had a chance to practice it two or three times.

Journal Questions

After students have completed the reading, give them the following questions for their reading journals, which will help them internalize the strategy they practiced:

- What did you do today that helped your reading group?
- Did you do anything that was unhelpful?
- What could you do next time to help your group even more?

Application/Post-Reading/Writing

- **Writing Prompt.** Use the questions at the end of your Racing for Control lab sheet (p. 32) to make a claim and support it with evidence and reasoning.
  - **Pre-Writing Suggestions.** Reasoning may be especially challenging in this lab because the data is very straightforward, and thus there is not much to explain. Help students see that explaining the ways in which they controlled the variables shows why this data supports the claim (since all the variables were the same except the race car, this indicates that the car itself must be the variable responsible for the win).
  - **Key Evaluation Point.** Students should make a reasonable claim based on their data. Their reasoning should show that valid scientific tests are about fairness. This includes keeping other variables the same to rule out other explanations.
- **Thinking Mathematically.** For this activity (pp. 38–42), students will use data to analyze car safety. It will be more effective and engaging if students access real data for cars of interest to them. If your students have internet access, direct them to *www.iihs.org/iihs/ratings*. Students will select three cars to compare (preferably within the same size category). Have them pull up the make and model of the first car, and give them a few minutes to click around and see the various safety ratings. Then help them find the data they need for comparison. (The steps for finding this information are on p. 42. Copy and pass out these instructions.) Note that older cars and the very newest cars may not have all crash test data. Students will need to pick a different car if they do not see data for the small overlap frontal crash test.* If your students do not have internet access, a worksheet filled with sample data has been provided, as well.

> **FIND OUT MORE**
> The Insurance Institute for Highway Safety has an excellent set of lessons that connects key physical science concepts with lab activities, videos, and safe driving information. Visit *https://classroom.iihs.org* for more information.

---

* These measurements were selected because they were easier than much of the other data for students to understand and analyze. Point out to your students that the data they are analyzing are not necessarily the most important data, and they should use the complete ratings page to make safety decisions about cars and car purchases.

Chapter 3

Fair on the Field—and in the Lab

Imagine that your team shows up for a basketball game. You walk into the gym and check out the basketball hoops. Your team is aiming for a typical hoop on a 10-foot pole. But the other team? Their hoop is waist-high and as big as a baby pool. Talk about unfair! Your coach wouldn't even let your team play the game under those conditions.

> **REMEMBER YOUR CODES**
> ! This is important.
> ✓ I knew that.
> X This is different from what I thought.
> ? I don't understand.

Every sport has an elaborate list of rules for keeping things fair. In track, runners start at different places along the oval race course depending on what lane they are in. Runners in the innermost lanes are behind those in the outermost ones. That way, the athletes in the outside lanes don't have a longer distance to run than the athletes in the inside lanes (see Figure S3.1). In soccer, teams trade sides halfway through the game in case one side of the field is easier to play on. In Olympic swimming, the two outer lanes of the pool aren't used, because those lanes have rougher waves that are more difficult to swim through.

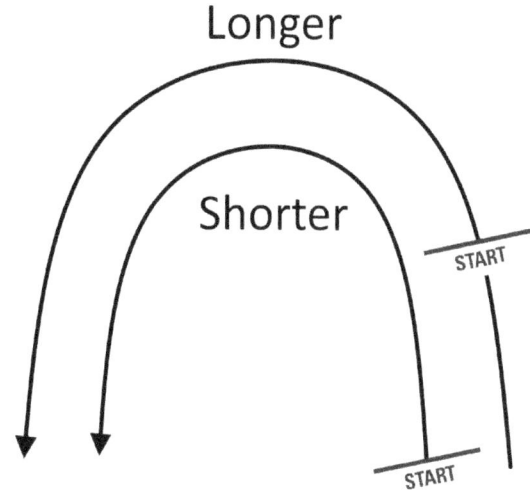

Figure S3.1. Outside and Inside Track Lanes

A controlled science experiment is similar to a sports event. When scientists design a controlled experiment, they put together "teams" and "rules." For example, imagine that a group of chemists are working to develop a new type of cell phone battery (see Figure S3.2). They want it to last a long time, even when it gets hot. They put their new batteries into 10 cell phones and see how long it takes the phones to die at room temperature (20°C). They also see what happens when the phones are placed in a hot car (50°C). The new batteries last just as long in both places. Sounds great, right?

Figure S3.2. Testing the Effect of Temperature on New Batteries

But what if some batteries were not fully charged before the test was started? It would be hard to compare how long the batteries lasted if they did not start at the same charge point. Or what if no apps were running on the hot phones, but the phones at room temperature were used for battery-draining games? If the same apps weren't used on the phones during the experiment, it wouldn't be a fair comparison.

The Smash-Masters

Researchers use the word *variable* to describe anything that can change, or vary, during an experiment. In the battery experiment, variables include how much power the batteries started with, what games or apps were played during the experiment, what kind of phone the batteries were in, and even whether the phones were old or new. To make a test fair, researchers want to keep everything the same between the groups except the variable they are testing. In this case, temperature should have been the only variable that was changed.

When a sports event starts, everyone is clear on what it takes to win. The winner will be the first athlete to cross the finish line, or the team with the most goals, or the team with the most runners to cross home plate. In the same way, researchers decide how they are going to compare their test groups. For the phone battery experiment, scientists were measuring how many hours it took each battery to die. At the end of the experiment, researchers use the data they gathered to make a claim. In the case of the phone battery experiment, the researchers could have claimed that the battery lasted longer at room temperature, the battery lasted longer in the hot car, or the battery performed about the same at both temperatures.

Controlled experiments are different from sports events in one big way. At a basketball game, the teams are looking to see who the best is in that particular game. Scientists are looking for patterns that happen over and over and over. If one battery does well at both temperatures, that's weak evidence to support the claim that this type of battery will always do well at both temperatures. If a thousand batteries do well, that's much better evidence. Scientists repeat experiments many times or with large groups to gather enough evidence to support their claim. In fact, they use a form of math called statistics to determine how many times they need to run their tests to get results they can trust.

When you are designing a controlled experiment, you can think like an athlete to design a fair test. Keep all the variables the same except for what you are testing. Decide what to measure so you will know which group did better. Do your test multiple times to develop strong evidence to support your claim.

Chapter 3

Racing for Control: Designing a Controlled Experiment

Question: Which car is faster?
Experimental Design: What will you do? (Provide your answers in the sections below.)
• Draw and label your experimental setup.
• What variables will you need to keep the same to have a fair test?
• How many times will you repeat the test? _____
• Specifically, what will you measure to compare the cars? (You may want to use terms like *how many*, *how far*, *how much*, or *how long*.)
• Make a data table to collect your results.

Once Upon a Physical Science Book

The Smash-Masters

Make a Claim:

Based on your data, how would you answer the question you asked at the beginning of this experiment?

Evidence:

Look at your data chart. Describe the data that support your claim. Use specific numbers from your chart.

Reasoning:

Explain how your data support your claim. Include how controlling the variables gives you trustworthy data.

Chapter 3

Reading-Group Roles

Your Role: Leader for Section 1

What to do in your group:
- Everyone should **read the section and code** (coding marks: ✓, X, ?, !).

- The leader for this round **tells what the section was about.** If you're stuck, try starting with, "What I understand so far is …"

- **Ask if anyone found something confusing** (marked X or ?).

  - The group should work together to figure out what the confusing word, sentence, or idea means.

  - If the group cannot make sense of it, raise your orange flag for help.

- **Turn to the next leader** and **repeat for the next section.**

- When your group has read and discussed all three sections, work together to **answer the Big Question.**

Your Role: Leader for Section 2

What to do in your group:
- Everyone should **read the section and code** (coding marks: ✓, X, ?, !).

- The leader for this round **tells what the section was about.** If you're stuck, try starting with, "What I understand so far is …"

- **Ask if anyone found something confusing** (marked X or ?).

  - The group should work together to figure out what the confusing word, sentence, or idea means.

  - If the group cannot make sense of it, raise your orange flag for help.

- **Turn to the next leader** and **repeat for the next section.**

- When your group has read and discussed all three sections, work together to **answer the Big Question.**

The Smash-Masters

Your Role: Leader for Section 3

What to do in your group:

- Everyone should **read the section and code** (coding marks: ✓, X, ?, !).

- The leader for this round **tells what the section was about.** If you're stuck, try starting with, "What I understand so far is ..."

- **Ask if anyone found something confusing** (marked X or ?).

  - The group should work together to figure out what the confusing word, sentence, or idea means.

  - If the group cannot make sense of it, raise your orange flag for help.

- **Turn to the next leader** and **repeat for the next section.**

- When your group has read and discussed all three sections, work together to **answer the Big Question.**

Chapter 3

The Smash-Masters

I watched, slack-jawed, as a car cruised by at a comfortable 40 miles per hour (64.4 kph). A few feet ahead, a large metal barrier blocked the road. Without even swerving, the driver plowed right into the barricade.

The "driver" was a dummy—a crash test dummy—and I was watching it in a video as it "did its job" for the Insurance Institute for Highway Safety (IIHS). Researchers at the IIHS perform dozens of crashes every year. The results of these crashes help shoppers select safer cars. They help insurance companies make decisions about insuring cars. And they help auto manufacturers know how to improve safety.

> REMEMBER YOUR CODES
> ! This is important.
> ✓ I knew that.
> X This is different from what I thought.
> ? I don't understand.

After seeing the video, I caught up with David Zuby, chief research officer for IIHS, to find out how they do the tests. He stressed that every vehicle is tested in exactly the same way and walked me through the process of doing a head-on test crash.

Getting Ready

"The first things we do are to take some pictures and measurements," Zuby explains. He weighs the car. He gets photographs of the car from several angles to show how it looked before the crash, and he measures what they call the "survival space." This is the amount of empty space available for the driver's body parts. Obviously, there's plenty of survival space before a crash. Hopefully, there's still a lot after the crash, as well.

The research team drains the gasoline, air conditioning coolant, and oil so they don't leave a big mess or catch fire when the car crashes. Then they start loading their test equipment.

The researchers add cameras, lights, systems that measure speed and forces, and a device that lets them put on the brakes if they need to stop the test. They even have a system that monitors whether all the other systems are working.

All this equipment is heavy. Since a car's weight strongly affects what happens in a crash, weight is an important variable for the researchers to control. Once the equipment is in place, the researchers weigh the car again. Then they remove items that won't affect how the car crashes (such as the spare tire) until they bring the weight back down to what that car would normally weigh on the road.

Next comes the crash test dummy (see Figure S3.3). The same style of dummy is used for all frontal crashes: It's 5'10" and

Figure S3.3. Crash Test Dummy

The Smash-Masters

weighs 172 pounds, which is about average for an adult male in the United States. The dummy is chock-full of sensors that can record what kind of impacts hit different parts of its body.

Of course, the dummy needs to have its seat and steering wheel adjusted for comfort. To decide what settings to use, researchers gathered a group of average-size males and had them climb into dozens of cars. They noted how those drivers adjusted the seat and steering wheel, and those adjustments are used to situate the crash test dummies. Before each test, the researchers record exactly how they place the dummy so that the test can be repeated if anyone needs to check the results.

"We're methodically controlling so that when the vehicle is set up for the test, it closely reflects the conditions of a car actually being driven by a regular person on the road," explains Zuby.

Crash-Smash-Bash

Finally, it's go-time. The car is connected to a crash machine that will propel it at exactly 40 miles per hour (64.4 kph). The researchers move away, activate the machine, and the car launches into the barrier—the same barrier that is used for every frontal test (see Figure S3.4). The entire crash is over in just a few seconds.

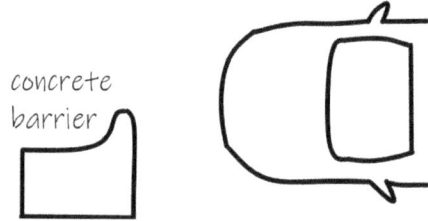

Figure S3.4. Setup for Head-On Crash Test

After the test, the researchers spring into action. They download the data from the dummy's sensors. These data will be compared to data from similar crashes involving real people to see what kind of injuries the dummy might have sustained. Then researchers remove the dummy from the car, documenting the steps they take. They want a record of how difficult it is to "rescue" the dummy after the crash. Afterward, they measure the survival space again. Although there may be enough space for the dummy to "survive" the crash, if the survival space is very small, a different-size dummy (or person) might not fare as well. The IIHS uses all the data from the crash to rate how well the car protected the driver.

Because the exact same process is repeated for each car the IIHS tests, consumers can check the results on the IIHS website before they buy a vehicle and pick one that is likely to keep them safe in an accident.

Using Models in Science

Of course, an actual car wreck might not happen in just the same way it did during the IIHS tests. The dummy in a crash is what scientists call a "model." Models are simplified versions of real life that scientists can use to understand an object or system. In this case, you wouldn't send a real person through a bunch of crashes to see if they get hurt. Therefore, scientists use the dummies. But models can't capture everything about a real person or system. For example, a crash test dummy has sensors in specific locations (such as the head, neck, and along the sides of the legs) to collect data for specific types of injuries. But when it comes to a real person, car crash injuries can occur anywhere on the body.

Figure S3.5. Real-Life Crashes Won't Match a Crash Test Exactly

Even the crash itself is a model for real-world crashes (see Figure S3.5). Scientists can't collect consistent before-and-after data on crashes that happen in real life. This is because they can't control the variables involved in a crash: the speed of the car, the weather and road surface conditions, and the size and angle of what the car hit. Therefore, they use the model crashes at the IIHS to create a controlled experiment where just one variable—the type of car—can be changed and compared.

The Big Question

List at least three variables that the IIHS keeps the same to make its tests fair and at least two types of measurements it collects.

Thinking Mathematically: Interpreting Data From Crash Test Dummies (Data Provided)

Most cars perform well in a crash test that models a head-on collision between two cars. But many cars are not as safe when there is only a small overlap between the tested car and what it hits. At the IIHS, this second situation is called a small-overlap frontal crash. The chart below shows some of the data collected from small-overlap crash tests.

Car	Category	Chest Compression (mm)	Force on Right Femur (kN)	Force on Left Femur (kN)
Car 1	Small car/4-door wagon	27	4.3	11
Car 2	Small car/4-door wagon	21	1.7	0.9
Car 3	Small car/4-door wagon	25	1.9	3.4

Note: A car's weight is very important in how a crash plays out. To keep their tests fair, IIHS only compares results for cars that are about the same weight. The IIHS assigns cars to categories based on weight. Within a category, you can compare safety results for the crash tests.

In a car crash, the seatbelt, the airbag, and the steering wheel can all squeeze the driver's chest. Studies with humans have shown that if the front of the rib cage is pushed in more than 50 mm (2 in), drivers are likely to have significant injuries to their ribs and internal organs.

1. Which driver had the most chest compression?

2. Did any of the dummies have enough chest compression to suggest serious injury? If so, which one(s)?

Crash test dummies record the force that hits their upper thigh, where a human has a large bone called the femur. If the force on the femur is at least 5.92 kN (or about as much force as a 1,330-pound weight on the leg), the bone is likely to break. (It also matters how *long* that force is hitting the bone. The IIHS takes that into consideration in its ratings. For simplicity, this worksheet ignores that part of the calculation.)

3. Which driver experienced the most force against the legs? Which leg was hit hardest?

4. Were any of the drivers hit hard enough to make a broken leg likely? If so, which one(s)?

5. If you were choosing a car for safety based on this information, which one would you pick? Use specific data to support your answer.

Thinking Mathematically: Interpreting Data From Crash Test Dummies (Online Data)

Most cars perform well in a crash test that models a head-on collision between two cars. But many cars are not as safe when there is only a small overlap between the tested car and what it hits. At the IIHS, this second situation is called a small-overlap frontal crash. Visit the IIHS ratings website at *www.iihs.org/iihs/ratings*. Locate a vehicle and model that you are interested in.

Once you have located your chosen car's web page on the IIHS website, your teacher will help you find the section titled "Small overlap front: driver-side," which includes technical measurements. Record the chest compression and forces on the femur in the chart below. Repeat for two other cars in the same category.

Car	Category	Chest Compression (mm)	Force on Right Femur (kN)	Force on Left Femur (kN)
Car 1				
Car 2				
Car 3				

Note: A car's weight is very important in how a crash plays out. To keep their tests fair, IIHS only compares results for cars that are about the same weight. The IIHS assigns cars to categories based on weight. Within a category, you can compare safety results for the crash tests.

In a car crash, the seatbelt, the airbag, and the steering wheel can all squeeze the driver's chest. Studies with humans have shown that if the front of the rib cage is pushed in more than 50 mm (2 in), drivers are likely to have significant injuries to their ribs and internal organs.

1. Which driver had the most chest compression?

2. Did any of the dummies have enough chest compression to suggest serious injury? If so, which one(s)?

Crash test dummies record the force that hits their upper thigh, where a human has a large bone called the femur. If the force on the femur is at least 5.92 kN (or about as much force as a 1,330-pound weight on the leg), the bone is likely to break. (It also matters how *long* that force is hitting the bone. The IIHS takes that into consideration in its ratings. For simplicity, this worksheet ignores that part of the calculation.)

3. Which driver experienced the most force against the legs? Which leg was hit hardest?

4. Were any of the drivers hit hard enough to make a broken leg likely? If so, which one(s)?

5. If you were choosing a car for safety based on this information, which one would you pick? Use specific data to support your answer.

Finding Data on the IIHS Website

1. Go to *www.iihs.org*, and click on "Vehicle Ratings" at the top of the page.

2. Enter the car make and model you would like to see. This will take you to a screen with the test results for the car you have selected.

3. Scroll down and look for a column on the left side of the page that lists tests. The IIHS sometimes runs different tests on different cars, so the tests listed here may vary from one car to another. However, one of the first tests listed should be "Small overlap front: driver-side." Click on this, then scroll down and click on "Technical measurements for this test." (*Note:* If the car you chose doesn't include the "Small overlap front: driver-side" test, pick another one to research.)

4. After selecting "Technical measurements for this test," find "Chest maximum compression" and "Femur" ("Left" and "Right"). Record the data you find.

Finding Data on the IIHS Website

1. Go to *www.iihs.org*, and click on "Vehicle Ratings" at the top of the page.

2. Enter the car make and model you would like to see. This will take you to a screen with the test results for the car you have selected.

3. Scroll down and look for a column on the left side of the page that lists tests. The IIHS sometimes runs different tests on different cars, so the tests listed here may vary from one car to another. However, one of the first tests listed should be "Small overlap front: driver-side." Click on this, then scroll down and click on "Technical measurements for this test." (*Note:* If the car you chose doesn't include the "Small overlap front: driver-side" test, pick another one to research.)

4. After selecting "Technical measurements for this test," find "Chest maximum compression" and "Femur" ("Left" and "Right"). Record the data you find.

Identify This

Chapter 4

Topics
- Physical properties
- Chemical properties
- Identifying unknowns

Reading Strategy
- Chunking

Identify This

Connections to Standards

Next Generation Science Standards (NGSS) Correlations	
Standard	
MS-PS1. Matter and Its Interactions (www.nextgenscience.org/dci-arrangement/ms-ps1-matter-and-its-interactions)	
Performance Expectation(s)	
The materials/lessons/activities outlined in this chapter are just one step toward reaching the performance expectation(s) listed below.	
MS-PS1-2. Analyze and interpret data on the properties of substances before and after the substances interact to determine if a chemical reaction has occurred.	

Dimension	Element	Matching Student Task or Question From the Activity
Science and Engineering Practice(s)	• Analyzing and Interpreting Data	• Students analyze and interpret data to provide evidence of whether a substance is the same or something new. • Students analyze and interpret data to determine similarities and differences in findings in the Thinking Mathematically section.
Disciplinary Core Idea(s)	**PS1.A.** Structure and Properties of Matter • Each pure substance has characteristic physical and chemical properties (for any bulk quantity under given conditions) that can be used to identify it.	• Students observe the physical and chemical properties of two initially similar liquids: vinegar and isopropyl alcohol. • Students learn to differentiate between physical and chemical properties by figuring out which tests change the substance and which did not. • Students read about more sophisticated techniques by law enforcement to identify unknown substances based on their properties.
Crosscutting Concept(s)	• Patterns	• Students observe macroscopic patterns as they explore the physical and chemical properties of two substances.
Common Core State Standards (CCSS) Correlations		
Reading Standard(s)	• CCSS.ELA-Literacy.RST.6-8.3. Follow precisely a multistep procedure when carrying out experiments, taking measurements, or performing technical tasks.	• Students conduct a series of chemical reactions using written instructions before designing their own procedure.
Writing Standard(s)	• CCSS.ELA-Literacy.WHST.6-8.1. Write arguments focused on discipline-specific content. • CCSS.ELA-Literacy.WHST.6-8.2. Write informative/explanatory texts, including the narration of historical events, scientific procedures/experiments, or technical processes.	• Students make a claim as to which test changed the substance being tested and support it with evidence from the lab. • Students explain which test would be best to perform on a substance first—one testing a physical property or one testing a chemical property.

Chapter 4

Background

Every pure substance is composed of groups of identical molecules, and these molecules will always have the same properties. This consistency is the basis of chemistry, and it allows scientists to identify unknown substances.

In this chapter, students will use physical and chemical properties to figure out the differences between some initially similar liquids. Then they will learn to differentiate between physical and chemical properties by figuring out which tests change the substance and which do not. They will also read about more sophisticated techniques that law enforcement uses to identify unknown substances in possible cases of illegal drugs.

As you teach this chapter, note that the word *property* is confusing for many students, including English language learners, because it means something different in everyday speech. Students may not have any word that they regularly use to describe the concept of a property, but you can get them started by helping them think of a property in science as being like a characteristic, trait, or feature. If these words are also unfamiliar, you can describe a property as "something that's always true about an object or thing."

Pre-Reading/Exploration

Materials for Activity

- Indirectly vented chemical-splash goggles
- Nitrile gloves
- Nonlatex apron
- Isopropyl alcohol (70%)
- Vinegar (5% acidity)
- Baking soda
- Table salt
- 1,000 ml beakers for ice baths (3–4 ice baths should work for the entire class)
- Ice (enough to fill the ice baths)
- Milk (refrigerated works best; one single-serve carton from cafeteria will do for entire class)
- Small paper clips (4–5 per group)
- Cotton balls (4–5 per group)
- Test tubes (3 per group)
- Blue litmus paper (4 per group, can be cut in half to save money)
- ¼ teaspoon measuring spoon for use with the baking soda (1 per group)

> **SAFETY NOTES**
> The following safety recommendations apply to all activities in this chapter:
> - Wear safety goggles, nitrile gloves, and nonlatex aprons during the setup, hands-on, and takedown segments of the activities.
> - Use caution when working with glass or plasticware, which can cut skin.
> - Never place food used in a lab activity in your mouth.
> - Review hazards on the Safety Data Sheets (SDSs).
> - Appropriately dispose of lab materials at the end of the activities as directed by the teacher.
> - Keep active flames away from flammable chemicals like alcohol as these can cause fire and/or explosions.
> - Use caution not to touch the saltwater ice bath. It is colder than typical ice water and can damage your skin.
> - Immediately report any lab accident to the teacher.
> - Wash your hands with soap and water after completing the activities.

> **TEACHING NOTE**
> If you need to reduce materials for this activity, you could remove the permanent marker test, the iron reaction, and/or the milk reaction. However, these are interesting reactions, so keep them if possible.

Identify This

- Small metric measuring cup or graduated cylinder (2–4 per group)
- Small cups (3 per group if reused for multiple tests)
- Petri dishes or saucers (2 per group)
- Dry erase board (1 per group or use classroom board for all)
- Black permanent marker (1 per group or shared among groups)

Activity

Use With Student Page(s): Same or Different? (lab sheet)

To introduce the lab, hold up a cup or glass of clear liquid. Tell students that you found it on your kitchen table and wondered what it was. Have them suggest various liquids it could be and ask them how they would go about deciding if it was a glass of water. Tell students that scientists have a number of ways to identify an unknown substance, and they are going to experiment with some of them today.

Part 1: Physical Properties

Students will analyze and test isopropyl alcohol and vinegar, following the instructions in the First Group of Properties section of the lab sheet. (See p. 50.) This section covers physical properties, although they will not be identified as such for the students just yet. The initial categories examined by students—color and state of matter—are intentionally similar so that you can explain that simple observations are often not enough to classify an unknown substance. When asked about color, students may have trouble settling on the idea of transparent or clear but encourage them to be as descriptive as they can.

The evaporation test is an easy way to get a sense of boiling point without complicated setup and safety requirements. The isopropyl alcohol will evaporate more quickly, and the test can be sped up by blowing over the top of both the alcohol and vinegar streaks at the same time.

You should prepare the saltwater ice baths ahead of time. (To create each ice bath, fill a 1,000 ml beaker with ice and add 10–15 ml of salt.) Freezing the vinegar should be quick with the addition of the salt, but you will want to be mindful that you are depressing the freezing temperature of the water. The isopropyl alcohol will not freeze.

Dissolving is sometimes a physical change and sometimes a chemical change. In the case of dissolved pigments, the change is physical. Permanent markers are not truly permanent but require the right solvent to remove the pigment. Isopropyl alcohol is one such solvent. Before giving the marker to students, test it in an inconspicuous place to make sure the alcohol will remove the brand you are using.

SAFETY NOTE
The saltwater solution is more dangerous than simple ice water. Students should not put exposed skin into the salt-and-ice mixture for any length of time.

Part 2: Chemical Properties

Students should start by setting up the reaction between the liquids and the paper clip. (For more information on this type of reaction, see Chapter 5). This is the slowest visible reaction, so it needs to begin quickly. By the end of the class period, the paper clip will feature a distinct dull-gray layer of reacted iron. However, bubbling can be seen throughout the reaction if students look closely. If you choose to leave the reaction to sit overnight, it will be darker and more distinct.

The second reaction involves litmus paper. In this case, we aren't trying to start a complex conversation about acids and bases. However, using the litmus paper as an indicator, students will be able to see a clear color change, and if they have discussed signs of a chemical reaction, they will recognize that color change is one of the signs. You can use both red and blue litmus paper if available, but only blue litmus paper is required. In order to minimize setup and cleanup, you can have the students dip their litmus test strips in the same cup they will use for the baking soda reaction.

For the baking soda reaction, choose a cup that has room for the foam from the reaction. The baking soda and vinegar reaction will be used to determine that a chemical property changes the substance. For this reason, it is important that the vinegar is completely reacted; in order to make this happen, students will be told to add a second ¼ tsp of baking soda. Students will then set the cup aside for the second part of the test.

The last reaction, which is with milk, works best if you have the students shake the test tubes before adding the vinegar and alcohol. The shaking separates the proteins and lipids so that it is easier to see the results. Once added, the vinegar reacts with the proteins to precipitate the casein. The alcohol might look to students like it is reacting, so pouring both the alcohol and vinegar solutions into petri dishes is an important final step. Students will see a smooth mixture with the alcohol, but distinct clumps will appear when the vinegar test tube is poured out into a petri dish.

Part 3: Did It Change?

Students will test the vinegar set aside after the physical property test on vinegar's freezing point and the chemical property test on vinegar's reaction with baking soda to figure out if a new substance was formed during these experiments. Students will be asked to use tests from both Parts 1 and 2 to make their determinations. Encourage students to pick tests that established distinct properties of the original substances. Only the chemical properties will give conclusive evidence: Litmus tests will no longer describe the vinegar as acidic and tests with milk and the paper clip will no longer change. If students decide to focus on the physical properties, they will not find the evidence conclusive: Evaporation will leave behind baking soda that was unreacted, the color is not particularly distinctive, and the freezing point will not be substantially different without much more careful measurement than students have already attempted.

Reading

Use With Student Page(s): "Was It a Drug Bust?" (article)

Introduce the Reading. Tell students they are going to read about a situation in which scientists regularly use physical and chemical properties to identify unknown substances.

Reading Strategy: Chunking

To introduce the strategy, display the following sentences:

Identical molecules of a chemical will always have the same properties. Molecules with an identical molecular structure will react the same way with other molecules, catch fire at the same temperature, and break apart in stomach acids at the same rate.

Point out that this is just two sentences, but they have a lot of ideas crammed in. Many sentences in science writing put a lot of information in a small space. It might be difficult to understand all the ideas at one time, but if students break the sentences into chunks, they can think about each piece individually. Now add slashes (/) to the sentences:

Identical molecules of a chemical/ will always have the same properties./ Molecules with an identical molecular structure/ will react the same way with other molecules,/ catch fire at the same temperature,/ and break apart in stomach acids at the same rate.

Talk students through each sentence, one section at a time. Start with the very first chunk: "Identical molecules of a chemical." Ask students to picture what that might mean. Consider drawing two water (H_2O) molecules on the board to show that their structure is the same. Follow with the next phrase, and remind students that they know what properties are because you have been discussing that in this unit. Then help them see that "molecules with an identical molecular structure" seems like a complicated phrase, but it means the same as "identical molecules of a chemical." Finally, walk them through the three properties listed (i.e., the identical molecules react the same way; catch fire at the same temperature [that is, they have the same combustion point]; and are digestible in stomach acids at the same rate) and point out that these are all properties they are already familiar with.

When you are finished, ask a student to summarize the information you have gathered from these two sentences. They may say something such as "When two things are the same molecule, they are the same thing. They have the same properties like the three properties listed at the end." Congratulate the students on pulling out the ideas from a difficult-sounding sentence.

Explain that chunking a sentence is like eating a pie. People cannot put a whole pie in their mouths at one time; they eat it bite by bite. When eating, some people will take smaller bites than others. Similarly, some people will need to break a sentence into more chunks than others, and that's okay. For this article, students can separate the chunks using slashes, like you did in your example. If they are reading from something they can't write on (i.e., a computer screen), they can chunk phrases in their heads or cover up parts of a sentence they aren't thinking about yet.

Journal Question

After students have completed the reading, give them the following question for their reading journals, which will help them internalize the strategy they practiced: Where did you use the chunking strategy in your reading today? Describe a phrase that was easier to figure out once you chunked it.

Application/Post-Reading/Writing

- **Writing Prompt.** Suppose you had only a small sample of an unknown. To help identify it, you wanted to test its density and see how it reacted with baking soda. Which property would you test first? Why? (Use the terms *physical property* and *chemical property* in your response.)
  - **Pre-Writing Suggestions.** Ask students what science words they could use in their responses. (Reinforce that they need to at least use the terms *physical properties* and *chemical properties*). Ask what writing words might be useful (e.g., *therefore* and *for example*).
  - **Key Evaluation Point.** Students should test density first because it is a physical property and the test will not change the substance. The baking soda test will be second because the reaction with baking soda is a chemical property and will change the substance.
- **Thinking Mathematically.** Using a table of properties in the Powder Confusion handout (p. 57), students will figure out what physical and chemical properties they can use to identify unknown powders.

> **FIND OUT MORE**
> Flammability is a distinctive chemical property of isopropyl alcohol, but the safety requirements of a demonstration are extensive. If you would like to show students this property, consider this video from Flinn Scientific: *www.youtube.com/watch?v=JM-trdzV1N4*.

Same or Different?

One way that scientists identify chemicals is by describing their properties. You are going to be given two chemicals. Use the instructions to create a description of each chemical.

SAFETY NOTE
Wear your protective goggles at all times. Do **NOT** smell or taste the liquids.

Part 1: First Group of Properties

A. Color: Write the color of each chemical in the First Group of Properties data table.

B. State of Matter: Is each chemical a solid, liquid, or gas at room temperature? Record your answers in the First Group of Properties data table.

C. Evaporation: Apply a small amount of vinegar to a cotton ball. Do the same with the alcohol. Run each cotton ball along your science table at the same time to create a THIN layer of liquid on the table. These should be equal amounts. Watch to see which substance evaporates (or disappears) first. Record what happens in the data table.

D. Freezing Point: The freezing point is the temperature at which a substance freezes. Do the alcohol and vinegar freeze at the same temperature? Place 1 ml of alcohol in one test tube and 1 ml of vinegar in another test tube. Your teacher will have prepared a saltwater ice bath. Set both test tubes into the ice bath at the same time.

SAFETY NOTE
Do **NOT** touch the saltwater ice bath at any point. It is colder than typical ice water and can damage your skin.

Check each minute and compare the substances until you see a difference. The one that freezes first has the higher freezing point. Remove the test tube of vinegar from the ice bath, **and set it aside for later use.**

E. Dissolving: Draw two lines on a dry erase board using a permanent marker. Using the cotton balls from the evaporation test, try to erase the marker. Did it erase with either one?

First Group of Properties

Substance	Color	State of Matter at Room Temperature (Solid, Liquid, or Gas)	Evaporation (Faster or Slower)	Freezing Point (Higher or Lower)	Dissolving (Erased Marker or Didn't Erase Marker)
Vinegar					
Isopropyl alcohol					

Chapter 4

Suppose you were given a substance, and you wanted to know if it was vinegar or alcohol. Which of the activities from Part 1 would help you find an answer?

Part 2: Second Group of Properties

A. Does it react with iron?
1. Measure out 10 ml of alcohol into in a small cup. Do the same in a second cup for vinegar.
2. Place a paper clip into the liquid with half of it above the liquid surface. It should lean against the side.
3. Look for changes to the paper clip throughout the remainder of the class period.
4. Record your observations in the Second Group of Properties data table.

B. Does it react with litmus?
1. Pour 10 ml of alcohol into a cup. Pour the same amount of vinegar into a second cup.
2. Dip half of the strip of litmus paper into each liquid and record your observations.

C. Does it react with baking soda?
1. Use the cups of liquid from the litmus test.
2. Add ¼ tsp baking soda to 10 ml of vinegar in a small cup. Record your observation in the data table.
3. Repeat the first step using isopropyl alcohol instead of vinegar.
4. Stir both solutions. To ensure that any chemical reaction is complete, add ¼ tsp additional baking soda. Stir. If any bubbling occurs, add another ¼ tsp baking soda.
5. Stir the vinegar and baking soda mixture **and set it aside for use in Part 3.**

D. Does it react with milk?
1. Place 5 ml milk in a clean test tube and label it "Vinegar." Cap the tube and shake it for three minutes to separate the lipids (fats) and proteins in the milk.
2. Add 2 ml vinegar.
3. Repeat Steps 1 and 2 using isopropyl alcohol instead of vinegar.
4. Swirl each test tube for 10 seconds and pour each solution into separate petri dishes or saucers.
5. Record your observations in the Second Group of Properties data table.

Second Group of Properties

Substance	React With Iron	React With Litmus	React With Baking Soda	React With Milk
Vinegar				
Isopropyl alcohol				

Suppose you were given a substance, and you wanted to know if it was vinegar or alcohol. Which of the activities from Part 2 would help you find an answer?

Part 3: Same or Different?

In Part 1, you set aside a test tube of vinegar that you previously froze. In Part 2, you set aside a cup of vinegar that you had mixed with baking soda. You are going to figure out if these containers still contain vinegar or if they have become something new.

Look at the charts for Parts 1 and 2. These will show you some of the properties of vinegar. Develop a plan to figure out if each container still has vinegar in it, or if something new has been formed. Prepare a data table for your results.

Your first step should be to pour the liquid from the vinegar and baking soda solution into a clean cup, leaving any baking soda that has settled in the old cup. That way, you have only liquid to work with.

Use only the tests you conducted in Part 1 or Part 2. Remember, you have a limited amount of liquid in each container. Choose your tests wisely! Use your results to answer the following questions:

A. Test Tube of Previously Frozen Vinegar

Claim: The test tube does/does not (circle one) still contain vinegar.

Evidence: (Describe results from the testing you did in Part 3 that support your claim.)

NATIONAL SCIENCE TEACHING ASSOCIATION

Reasoning: (Use the data from Parts 1 and 2 to explain why the results of your evidence support your claim.)

B. Cup of Vinegar Mixed With Baking Soda

Claim: The cup does/does not (circle one) still contain vinegar.

Evidence: (Describe results from the testing you did in Part 3 that support your claim.)

Reasoning: (Use the data from Parts 1 and 2 to explain why the results of your evidence support your claim.)

Was It a Drug Bust?

A police officer knocked on the door of the house and presented a search warrant. Other officers surrounded the house and arrested two suspects as they tried to flee. Inside, the police found bag after bag of white powder.

The officers suspected that the bags contained methamphetamine, called *meth* for short. Meth is a dangerous drug that gives users a temporary high but damages every part of the body that it touches. For a drug, meth is a simple molecule. It has 10 carbon atoms, 15 hydrogen atoms, and 1 nitrogen atom, as shown in Figure S4.1. If a sample of meth is "pure," then all the molecules in the sample will have this exact same structure.

One of the suspects immediately claimed that the bags contained ground-up aspirin. He said that he ground the aspirin to make pastes for his grandmother's aches and pains. How could the officers find out if the suspect was telling the truth?

> REMEMBER YOUR CODES
> ! This is important.
> ✓ I knew that.
> X This is different from what I thought.
> ? I don't understand.

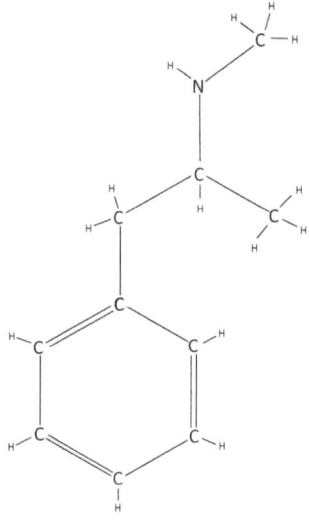

Figure S4.1. A Molecule of Meth

On-the-Spot Testing

The officers began their analysis with a testing kit they had brought with them. Their test involved mixing a few drops of chemicals from the kit with the unknown substance. As you can see at the top of Figure S4.1, meth has a nitrogen (N) sandwiched between two carbons (C). Any molecule with that shape would react in this test to create a new, bright-orange molecule. Aspirin does not have nitrogen and won't turn orange.

This test works because identical molecules of a chemical will always have the same properties. Molecules with identical molecular structure will react the same way with other molecules, catch fire at the same temperature, and break apart in stomach acids at the same rate. If a molecule goes through any of these processes, the molecule changes. Properties that can only be discovered by changing the molecule are called chemical properties.

To the Lab

Unfortunately for the suspect, the powder turned bright orange when it was tested, showing that at least part of the molecule was the right shape to be meth. The police sent another sample to a drug lab for more testing.

Most illegal drugs are not pure. They may contain several different chemicals mixed together, some of which can be even more dangerous than the drug being purchased. The first step in a drug lab is to separate the different chemicals in the sample.

A technician at the lab fed the sample sent by the police into a machine called a gas chromatograph, or GC for short. When a sample is placed in a GC, it is first heated until it reaches its boiling point and becomes a gas. Then the gas floats through a long tube that is about the length of three school buses. The tube is twisted tightly so it will fit in the machine. As shown in Figure S4.2, molecules move through the tube at different speeds. Because the tube is so long, different molecules reach the end at different times. Small molecules with a low boiling point tend to move through the fastest, whereas large molecules with a high boiling point move more slowly.

Molecule size and boiling point are physical properties, or properties that can be tested without changing the molecules. Therefore, the GC sorts the chemicals by their physical properties.

Figure S4.2. Molecules Move Through the GC at Different Speeds

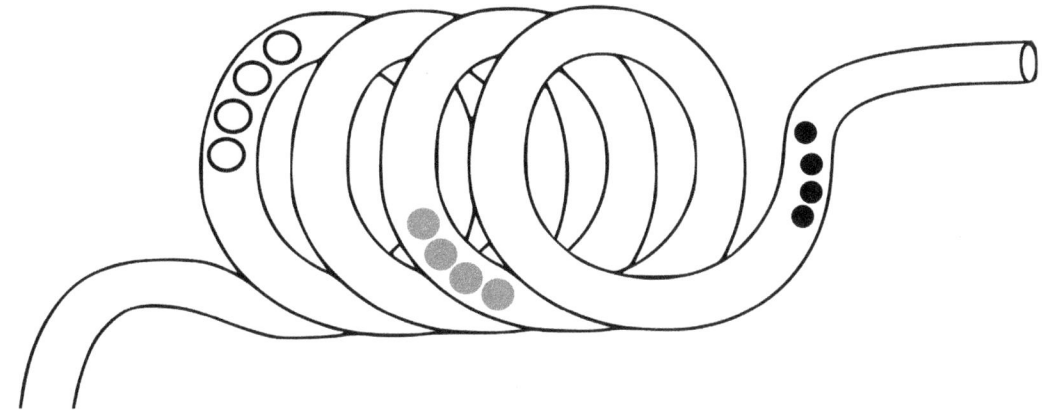

Researchers in drug labs know how long it should take different molecules to travel through their machines. For example, cocaine is a large molecule. It takes 13.5 minutes to travel through certain machines. On the same machines, meth zips through in just 5.1 minutes.

The powder from the crime scene? It popped out in 5.1 minutes. This result indicated that the chemical might indeed be meth. But the identification process was not complete because closely related molecules can travel through the GC at similar times, so one more step was needed to identify the powder.

Identifying White Powder

After GC analysis, molecules from a sample move straight from the GC into another machine called a mass spectrograph, or MS. An MS breaks the molecules into chunks, as shown in Figure S4.3 (p. 56). Molecules break up into chunks the same way every time. This is another chemical property because it definitely changes the molecules involved. Once again, the chunks travel through a long tube.

Identify This

As the chunks exit the MS, the machine records their masses. It prints a graph of how many chunks of each size were in the sample. If the sample is meth, the biggest chunks will have a molecular mass of 134, 91, and 58.

The technician working on the possible meth sample checked the data as the sample rolled through the MS. A few minutes later, she had the readout of the molecular mass: 134, 91, and 58.

She marked the sample "Methamphetamine" and carefully filled out her paperwork. This was nobody's grandma's aspirin, and this evidence would make sure the police could prove it in court.

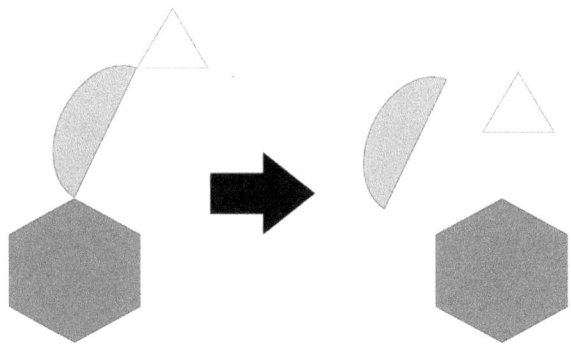

Figure S4.3. The MS Breaks Molecules Into Chunks

The Big Question

Look back at your lab. You tested two groups of properties. Which group contained physical properties? Which set were chemical properties? How do you know?

Chapter 4

Thinking Mathematically: Powder Confusion

You work for a shipping company, and the labels have fallen off a shipment of white powders. Look at the properties in the Shipping List Chart and decide which are physical and which are chemical. Then use the information to decide how to get the labels back on the right containers. Use the chart and the list of information that follows to decide which substance is in each jar:

Shipping List Chart

	Melting Point (°C)	Density (g/cm^3)	Appearance	Reaction With Bromothymol Blue	Reacts With Air?
Physical or Chemical Property?	_____	_____	_____	_____	_____
Table salt (NaCl)	801	2.17	White crystals	Turns green	No
Baking soda (Sodium Bicarbonate)	50	2.20	White powder	Turns blue	Reacts with moisture in air
Sugar (sucrose)	185	1.6	White crystals	Turns green	No
Explosive (ammonium nitrate)	170	1.7	White crystals	Turns yellow	No
Lye (sodium hydroxide)	323	2.13	White crystals	Turns blue	Reacts with CO$_2$ in air

1 _____   2 _____   3 _____   4 _____   5 _____

- Jar 1 and Jar 2 are both specially sealed to keep air out. Which two substances should be in those jars?

- You don't want to accidently heat the jar with the explosive. So you need to identify that jar before you start testing melting points. What test should you do to identify the explosive?

- A sample from Jar 4 turns yellow when mixed with bromothymol blue. Samples from Jar 3 and Jar 5 both turn green.

- When you put them both over a Bunsen burner, a sample from Jar 1 melts before a sample from Jar 2.

- A sample from Jar 3 has a higher density than a sample from Jar 5.

Handy Heaters

Chapter 5

Topics

- Chemical reactions
- Chemical equations
- Chemical hand warmers

Reading Strategy

- Reading technical text

Connections to Standards

\multicolumn{3}{	c	}{Next Generation Science Standards (NGSS) Correlations}

Standard
MS-PS1. Matter and Its Interactions (*www.nextgenscience.org/dci-arrangement/ms-ps1-matter-and-its-interactions*)

Performance Expectation(s)
The materials/lessons/activities outlined in this chapter are just one step toward reaching the performance expectation(s) listed below.

MS-PS1-2. Analyze and interpret data on the properties of substances before and after the substances interact to determine if a chemical reaction has occurred.

MS-PS1-5. Develop and use a model to describe how the total number of atoms does not change in a chemical reaction and thus mass is conserved.

Dimension	Element	Matching Student Task or Question From the Activity
Science and Engineering Practice(s)	• Developing and Using Models	• Students use Ping-Pong balls and egg crates to describe the chemical reaction of iron and oxygen at the molecular level. • Students use symbolic models in the Thinking Mathematically section to illustrate that the total number of atoms in a chemical reaction does not change.
Disciplinary Core Idea(s)	**PS1.A.** Structure and Properties of Matter • Each pure substance has characteristic physical and chemical properties (for any bulk quantity under given conditions) that can be used to identify it. **PS1.B.** Chemical Reactions • Substances react chemically in characteristic ways. In a chemical process, the atoms that make up the original substances are regrouped into different molecules, and these new substances have different properties from those of the reactants. • The total number of each type of atom is conserved, and thus the mass does not change.	• Students observe the action of a disposable hand warmer and set up a chemical reaction to create a modified version of a hand warmer. • Students use Ping-Pong balls as a model to illustrate the regrouping of atoms in the original substance into different molecules in the new substance. • Students observe the difference between a fully reacted model hand warmer (unsealed) and a sealed model where the reaction was stopped due to an insufficient amount of oxygen (reactant) available in the chemical reaction. • Students use symbolic models in the Thinking Mathematically section to illustrate law of conservation of mass. • Students read about how hand warmers release heat when they undergo a chemical reaction and how to balance the chemical equation according to law of conservation of matter.
Crosscutting Concept(s)	• Patterns • Energy and Matter	• Students observe the macroscopic pattern of iron oxidation when they set up a chemical reaction to create a modified version of a hand warmer and use a Ping-Pong model to relate it to the atomic-level (microscopic) structures of the reaction. • Students use a Ping-Pong ball model and the Thinking Mathematically section to illustrate the conservation of matter in a chemical process.

Continued

Chapter 5

Common Core State Standards (CCSS) Correlations		
Reading Standard(s)	• CCSS.ELA-Literacy.RST.6-8.7. Integrate quantitative or technical information expressed in words in a text with a version of that information expressed visually (e.g., in a flowchart, diagram, model, graph, or table).	• The reading strategy for this chapter asks students to practice integrating chemical reactions represented in symbols with verbal representations in the text.
Writing Standard(s)	• CCSS.ELA-Literacy.WHST.6-8.2. Write informative/explanatory texts, including the narration of historical events, scientific procedures/experiments, or technical processes. • CCSS.ELA-Literacy.WHST.6-8.2.a. Introduce a topic clearly, previewing what is to follow; organize ideas, concepts, and information into broader categories as appropriate to achieving purpose; include formatting (e.g., headings), graphics (e.g., charts, tables), and multimedia when useful to aiding comprehension.	• Students use what they have learned from reading and experimenting to write a package insert for hand warmers that explains the science concepts involved in the heat production. • Students are specifically asked to integrate a diagram or other graphic into their explanation.

Background

In this chapter, students will observe the action of a disposable hand warmer, set up a chemical reaction to create a modified version of the hand warmer, and then use models to think about what is happening at the molecular level.

Chemical reactions can be difficult for students to conceptualize because most of the action takes place at a scale too small to observe. Throughout this chapter, help students realize that they are looking at the same reaction at the macro level (as they make the hand warmers), at the atomic level (when they model what's happening with Ping-Pong balls), and at the symbolic level (when they are describing the reaction in the form of a chemical equation).

Before using this chapter, students should be familiar with physical and chemical changes and understand that elements and molecules can be represented using the symbols on the periodic table.

Pre-Reading/Exploration

Materials for Activity (Per Group)

- Indirectly vented chemical-splash goggles
- Nitrile gloves
- Nonlatex aprons

SAFETY NOTES
The following safety recommendations apply to all activities in this chapter:
- Wear safety goggles, nitrile gloves, and nonlatex aprons during the setup, hands-on, and takedown segments of the activities.
- Use caution when working with glass or plasticware, which can cut skin.
- Review hazards on the Safety Data Sheets (SDSs).
- Use caution when handling steel wool, which can puncture or scratch skin.
- Appropriately dispose of lab materials at the end of the activities as directed by the teacher.
- Immediately report any lab accident to the teacher.
- Wash your hands with soap and water after completing the activities.

Once Upon a Physical Science Book

Handy Heaters

- 1 disposable hand warmer (Buy hand warmers in bulk online to save money.)
- 1 sandwich bag
- Steel wool pads cut into halves, one half-pad per group
- 200 ml of vinegar in beaker or bowl
- 1 thermometer (nonmercury)
- Empty egg cartons or pieces cut from a twin eggcrate foam mattress pad (30 egg slots per group)
- 14 Ping-Pong balls (6 labeled "O," 4 labeled "H," and 4 labeled "Fe")
- 1 cardboard arrow

Activity

Use With Student Page(s): Handy Heaters (lab sheet)

Before class, prepare the Ping-Pong balls and egg cartons for each group. Each group will need:

- 2 *pairs* of egg slots labeled "H" inside with Ping-Pong balls also labeled "H" to represent two hydrogen molecules
- 3 *pairs* of egg slots labeled "O" inside with Ping-Pong balls also labeled "O" to represent three oxygen molecules
- 2 empty *triads* of egg slots labeled "H-O-H" to represent water (see Figure 5.1)
- 4 *single* egg slots labeled "Fe" with Ping-Pong balls also labeled "Fe" to represent atoms of iron
- 2 empty *groups of 5* egg slots labeled "O-Fe-O-Fe-O" (as shown in Figure 5.2, p. 64) to represent molecules of rust

Gathering and preparing these materials can be time consuming, but the materials can be used over and over in the years to come. You may also wish to divide each of your steel wool pads into two pieces to help the supplies go further.

In Part 1 of the activity, students will observe store-bought chemical hand warmers. It takes about two minutes for the hand warmers to get warm. During this time, students will brainstorm their prior knowledge about rusting. For the first minute, they will record ways to tell that a bicycle is old. For the second minute, you will discuss as a class which changes to the bicycle would be physical and which would be chemical. Students may not know whether rusting is a physical change or a chemical one. Do not tell them one way or another, but revisit the question after the lab.

In Part 2, students will create a modified hand warmer using vinegar and steel wool. Note that the lab sheet refers to the wool as "iron wool" to emphasize the importance of iron content in the creation of the hand warmers. If students ask, you can explain that steel wool is primarily (up to 95%) iron. At the conclusion of this activity, instruct half the class to press out the air from their bags and to seal them shut. Leave the other half of the bags open to the air overnight.

To begin Part 3, tell students that they have just seen a chemical reaction. Walk them through what went into the reaction (i.e., iron and air). Ask students to describe signs that a reaction occurred. (Answer: The reaction resulted in heat and rust.) Say that with a chemical reaction, we cannot see what is happening to the individual molecules. But we can use a model to understand what is happening. You will probably need to walk students through this activity, discussing what is happening as students work with the models.

For the first example, they will use hydrogen and oxygen egg carton molecules to show a simple example of a chemical equation: the formation of water from hydrogen and oxygen molecules. Give each group of students the materials shown in Figure 5.1. Ask them to take the Ping-Pong ball atoms from a hydrogen molecule (H_2) and an oxygen molecule (O_2) and make a water molecule (H_2O) using the empty egg carton frame.

Figure 5.1. Water

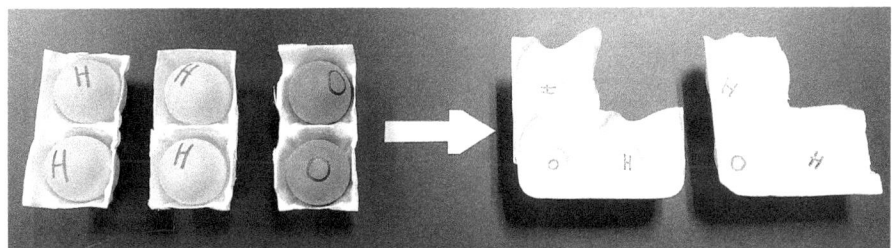

They will discover that they have one extra oxygen atom when a water molecule is made. Have them place the extra oxygen atom into a second water molecule frame. Ask the students how much hydrogen is needed to complete the second water molecule. (*Note:* They will need one more hydrogen molecule [H_2] to complete the second water molecule.)

Show students that they can represent what they have done in writing, using the following chemical equation: $2H_2 + O_2 \rightarrow 2H_2O$. Explain that what they started with is on the left side of the equation and what they made is on the right.

Remind students of the law of conservation of mass and point out that a balanced equation has the same number of each type of atom on both sides. Make a T-chart to demonstrate:

Left Side of Equation	Right Side of Equation
4 hydrogen (H)	4 hydrogen (H)
2 oxygen (O)	2 oxygen (O)

Give students the materials shown in Figure 5.2. Have them use the same process shown in the figure to complete the chemical equation for iron oxidation, or rusting. Note that this lab simplifies the chemical processes involved in rusting. There are several steps to the rusting process and more than one compound that can potentially be produced. Here, we are using the most straightforward series of reactions and moving straight from initial reactants to final products, leaving out intermediary molecules.

Figure 5.2. Iron Oxide

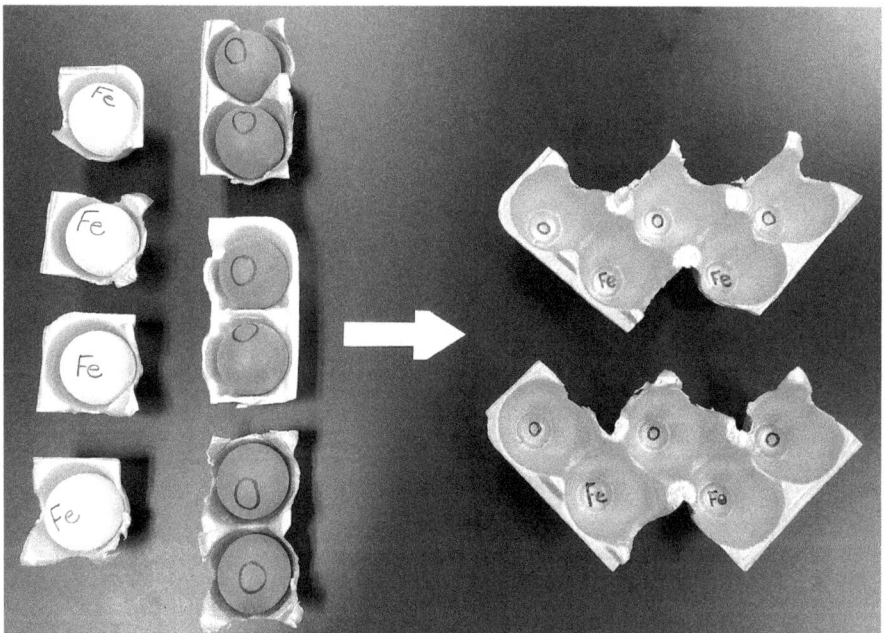

Students will start with two Fe atoms and one O_2 molecule. They will need one more oxygen atom from another O_2 molecule for iron oxide. The remaining O from the molecule will start a new iron oxide molecule. They will have to determine how many more atoms they need to complete the second iron oxide frame. The equation for the reaction is as follows: $4Fe + 3O_2 \rightarrow 2Fe_2O_3$. Have students fill in a T-chart and observe that the numbers of atoms on each side are equal.

Left Side of Equation	Right Side of Equation
4 iron (Fe)	4 iron (Fe)
6 oxygen (O)	6 oxygen (O)

On the day after this lab, students will complete Part 4, the final section of the activity. Have students revisit the bags that were sealed and those that were not. The sealed bags will have run out of oxygen, stopping the reaction. The bags left open will have continued to react. Students will see that the iron in the open bags has rusted through and will crumble when squeezed. Remind students that a chemical reaction requires all the reactants to be present. They can model this using the Ping-Pong balls; students will see that they cannot continue making rust if they do not have any oxygen molecules. Show students that you can reinitiate the reaction in the sealed bags by opening them. Connect this back to the store-bought hand warmers: Sealing hand warmers in a resealable bag allows you to save a partially used hand warmer for another day.

Reading

Use With Student Page(s): "Frostbite Free" (article)

Introduce the Reading. Tell students they are going to read more about how hand warmers work and why they are useful.

Reading Strategy: Reading Technical Text

Reading text that includes chemical equations requires a different reading pattern to that of traditional text. Readers need to work through what is being said about an equation as it is described. In this book, the term *back-and-forth reading* is used to describe a strategy for reading equations. To introduce the strategy, display this text:

> *Chemical equations can be used to describe what is happening in a chemical reaction. The reaction for creating rust might be written as follows: $Fe + O_2 \rightarrow Fe_2O_3$.*
>
> *In this equation, iron and oxygen are the starting materials, called reactants. Fe_2O_3, the new molecule of rust, is called the product. Notice that as a reactant, oxygen appears as O_2. Oxygen atoms do not hang out by themselves. An individual oxygen atom will join another oxygen atom to form one molecule. Thus, oxygen gas is written as O_2 in equations to show that the oxygen is present in pairs. There is a problem with this equation, however.*

Handy Heaters

Tell students that reading text with equations in it can be tricky because the equations seem like a jumble of letters, and many times what matters about the equation isn't immediately obvious. It's not until you read the words just before or after the equation that you understand what the author wants you to get from it. This means you have to do some back-and-forth reading, where you read the words and keep checking the equation to see what the author is talking about.

Read the first sentence out loud. Comment that the text lets you know that you are dealing with an equation for rust. Set your finger by the equation to make it easy to look back and forth as you read the text that follows.

Read the sentence following the equation out loud: "In this equation, iron and oxygen are the starting materials, called reactants." Ask students to look back at the equation to find the iron and oxygen. Note their location and that they are called reactants. Jot the word *Reactants* above the first part of the equation.

Read the next sentence: "Fe_2O_3, the new molecule of rust, is called the product." Ask students what you should do next. (Answer: Look back to the equation and find Fe_2O_3.) Label the Fe_2O_3 molecule in the equation as "rust" and write the word *Product* above the second part of the equation. Read the next few sentences about oxygen existing in pairs and ask what a reader should do after reading those statements. (Answer: Look at the equation to see how the oxygen molecule is written.) With the final line "There is a problem with this equation, however," point out that students will want to be watching for what that problem is as they read further. Tell students to continue with back-and-forth reading as they read the article "Frostbite Free."

Journal Question

After students have completed the reading, give them the following question for their reading journals, which will help them internalize the strategy they practiced: Think about the strategy of back-and-forth reading with chemical equations. How would you explain it to a friend who was absent today?

Application/Post-Reading/Writing

- **Writing Prompt.** Write an insert for a hand warmer package to explain to users how it works. Include at least one equation and one diagram in your answer.
    - **Pre-Writing Suggestions.** Ask students what science words would be useful in their inserts. (Specify for students if you

expect certain words.) Ask them what equation(s) they would want to include. Have students think about where they can look for information if they aren't sure what to write. Ask why showing a diagram in the insert would be helpful to the reader. Remind students to refer to their diagram and equation in the text, using a sentence starter such as "As you can see in the diagram and equation …"
    - **Key Evaluation Point.** Students should explain that rusting is a chemical reaction that releases heat energy. Furthermore, they should include the equation for rusting. The most appropriate diagram would show atoms interacting to make iron oxide. However, students may generate other diagrams, as well.
- **Thinking Mathematically.** Using the Balancing Chemical Reactions worksheet (p. 74), students will examine visual representations of reactions to balance chemical equations.

Handy Heaters

Part 1: Examining Hand Warmers

Tear open one of the packets that your teacher has provided. Gently shake your packet for a few seconds. Set it on the corner of your desk.

1. Imagine an old bicycle that has been sitting outside for many years. Describe how the bike might look.

2. Of the descriptions you gave in the first question, which refer to physical changes to the bike? Which refer to chemical changes?

3. Now reach for the packet. Describe how it feels.

4. Take a guess about what you think happened in the packet.

Part 2: Make Your Own Hand Warmer

1. Test the temperature of your iron wool pad by nestling the thermometer in it and placing it in your bag. Leave it until the temperature on the thermometer is stable. Write the starting temperature of your iron wool in the Temperature Reading Chart.
2. Pour 200 ml of vinegar into a bowl.
3. To get the reaction started, you'll need to use vinegar to remove a protective coating that manufacturers apply to iron wool. Remove the iron wool pad from the bag and dunk it completely into the vinegar in the bowl. You might have to turn the pad over to make sure all its surfaces have been touched by the vinegar.
4. Wring out as much vinegar from the iron wool as you can.

5. Place your iron wool pad back into the plastic bag and place the thermometer into the middle of the wool. Seal up the bag with a little air inside. It is okay if your thermometer is sticking out of the bag. Just make sure you can see the temperature reading.

6. Watch the iron wool pad for the next few minutes, recording its temperature in the Temperature Reading Chart.

Temperature Reading Chart

Starting temperature	
After 30 seconds	
After 1 minute	
After 1 minute 30 seconds	
After 2 minutes	
After 3 minutes	
After 5 minutes	

7. As you watch the iron wool and record its temperature, what other observations can you make about what is happening inside the bag?

8. Your teacher will direct you to either seal your bag or leave it open. If you are sealing your bag, press out as much air as possible before sealing it. Leave the bags overnight.

Part 3: Reaction Modeling

1. Complete the chemical equation for water and corresponding T-chart with your teacher's guidance.

_____ + _____ → _____

Left Side of Equation	Right Side of Equation

2. This time, model the chemical reaction for rusting, also called iron oxidation. Start with two Fe atoms and one O2 molecule. (Oxygen atoms hang out in pairs.)

   a. How many more oxygen atoms are needed to complete the iron oxide molecule?

   b. What additional atoms are needed to complete the second iron oxide frame?

   c. What is the chemical equation for iron oxidation?

   _____ + _____ $\rightarrow$ _____

   d. Check your equation by filling out the T-chart. Do you have equal numbers of atoms on each side?

Left Side of Equation	Right Side of Equation

Part 4: Oxygen on Hold

1. Think about the rusting reaction you just modeled and the hand warmers you made earlier. Predict what you think will happen to the bags that are remaining open and exposed to oxygen overnight.

2. What do you predict will happen to the bags that are sealed?

Chapter 5

Frostbite Free

Every year, close to 100 dog-sled racers, known as mushers, compete in a wild race across Alaska (see Figure S5.1). In this race, called the Iditarod, mushers travel from the city of Anchorage to the city of Nome, driving a sled pulled by a team of dogs. It's a grueling journey of nearly 1,000 miles. The racers and their dogs trek night and day, grabbing only a few hours of sleep when the dogs need rest. What's more, they travel through blizzards during which the temperature can drop as low as −60°F (−51°C).

REMEMBER YOUR CODES
! This is important.
✓ I knew that.
X This is different from what I thought.
? I don't understand.

The biggest danger on the trail is frostbite. With frostbite, sections of fingers, toes, or other body parts freeze and die. The frozen portions can fall off on their own or may need to be amputated. In the temperatures racers face on the Iditarod, exposed body parts can develop frostbite in less than five minutes. Gloves and socks are critical for keeping warm but are not always enough. When mushers worry that their hands or feet are getting too cold, they reach for chemical hand warmers.

Figure S5.1. Racing With Sled Dogs

Making Heat in the Cold

Hand warmers seem like magic, pulling heat from nowhere. In reality, though, these items release heat as they undergo a chemical reaction. A chemical reaction takes place when atoms or molecules change or rearrange to make something new. The new type of molecule made inside a hand warmer is rust, the same brown-orange stuff that shows up on old cars, old nails, or bicycles that have been left out in the rain. Chemically, rust is written as Fe_2O_3 to show that each molecule contains two atoms of iron (Fe) and three atoms of oxygen (O) joined together.

Chemical equations can be used to describe what is happening in a chemical reaction. The reaction for creating rust might be written as follows:

$$Fe + O_2 \rightarrow Fe_2O_3$$

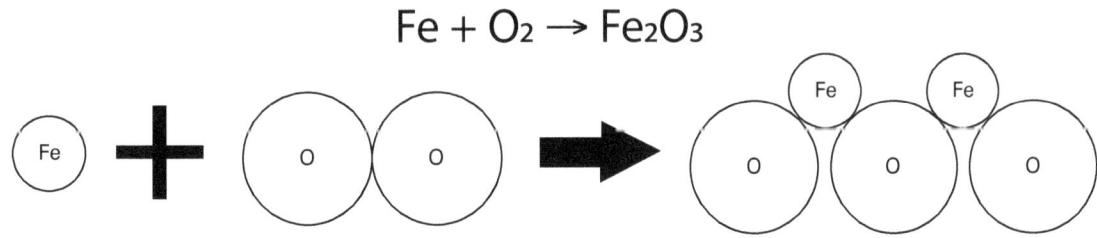

In this equation, iron and oxygen are the starting materials, called reactants. Fe2O3, the new molecule of rust, is called the product. Notice that as a reactant, oxygen appears as O2. Oxygen atoms do not hang out by themselves. An individual oxygen atom will join another oxygen atom to form one molecule. Thus, oxygen gas is written as O2 in equations to show that the oxygen is present in pairs.

Out of Balance

There is a problem with this equation, however. The reactants in the equation have *two* oxygen atoms, but there are *three* oxygen atoms in the product. The iron has a similar problem. The law of conservation of matter states that matter cannot be created or destroyed—so where did the extra atoms come from? The extra iron and oxygen didn't spring from nothingness. Look at the corrected equation that follows:

$$4Fe + 3O_2 \rightarrow 2Fe_2O_3$$

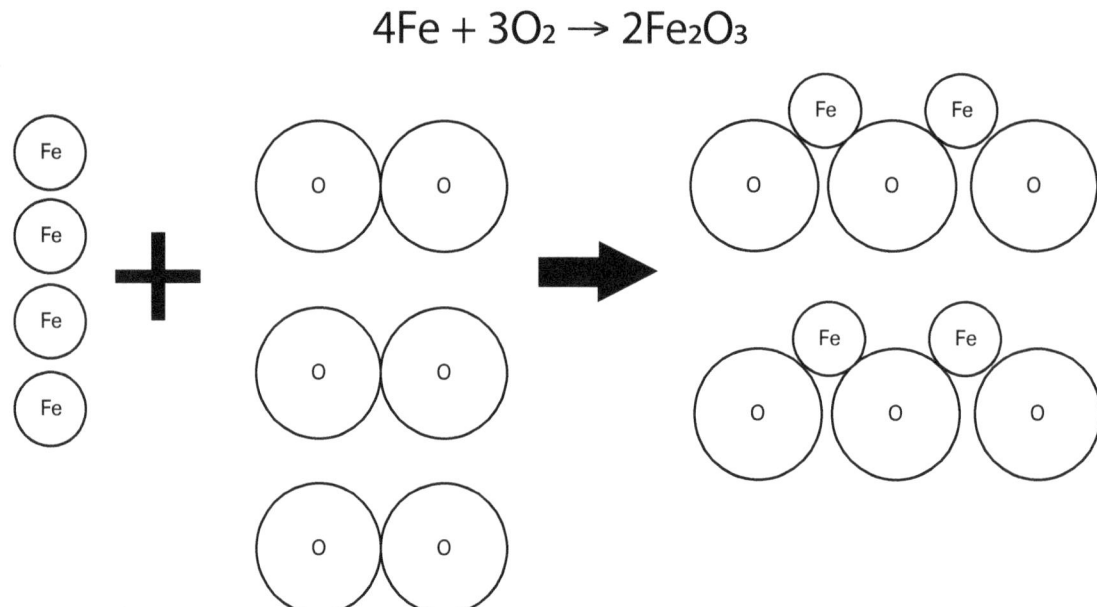

When iron rusts, *groups* of iron atoms react with *groups* of oxygen molecules to form *groups* of rust molecules. A balanced chemical equation shows how many groups of each are needed to come out evenly. In the equation above, there are four iron atoms combining with three pairs of oxygen atoms to make two full molecules of rust. There are four iron atoms and six oxygen atoms on both sides of the equation, so it is balanced.

NATIONAL SCIENCE TEACHING ASSOCIATION

Getting Warmer

Sometimes an equation needs to show more than just the molecules involved. It may need to show something about the conditions in which the reaction takes place or other kinds of changes that occur during the reaction. In the case of rusting, the chemical reaction also releases thermal energy. The thermal energy can be added into the equation like this:

$$4Fe + 3O_2 \rightarrow 2Fe_2O_3 + \text{thermal energy}$$

We feel the release of thermal energy as heat. On a dog sled, that heat can be the last line of defense against frostbite.

Save Some for Later

Sometimes a musher may need just a little burst of heat. She may not want to use up her entire hand warmer. How can she stop the reaction and save some of her iron for later? Look back at the equation for rusting. Notice that the reaction requires two reactants: iron and oxygen. If either one is missing, the reaction cannot happen. If the hand warmer has a constant supply of oxygen, all the iron will eventually turn to rust. If the musher wants to stop the reaction sooner, all she has to do is seal the hand warmer in an oxygen-free container such as a tight plastic bag. No oxygen means no rusting. Those iron filings sit tight until the musher is once again feeling desperate for some heat.

The Big Question

Table salt (NaCl) is made from sodium (Na) and chloride (Cl_2). Like oxygen, chloride atoms stick together in pairs. Write an equation for adding sodium and chloride to make table salt. Is your equation balanced? How do you know?

Thinking Mathematically: Balancing Chemical Reactions

Equation 1: Water

This diagram shows a chemical reaction.

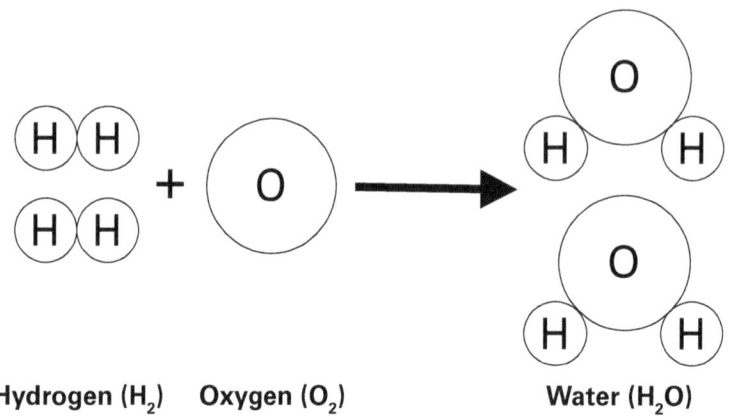

1. What are the reactants?

2. What are the products?

3. Is this equation balanced? (That is, are there the same number of each type of atom on both sides of the equation?)

Equation 2: Methane and Oxygen

When methane reacts with oxygen, carbon dioxide and water are produced, as shown on the following page.

Chapter 5

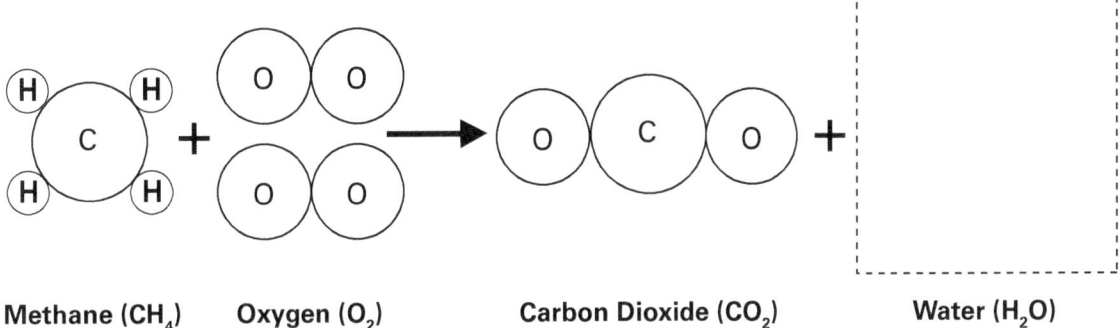

Methane (CH$_4$)   Oxygen (O$_2$)   Carbon Dioxide (CO$_2$)   Water (H$_2$O)

4. Draw atoms in the space provided to add the missing water diagram to the chemical equation.
5. Is the chemical equation balanced? (If not, draw more water molecules to balance it.)

Equation 3: Photosynthesis

Plants and animals greatly benefit from chemical reactions. In one important reaction, plants use carbon dioxide, water, and energy from the Sun to make oxygen and a sugar called glucose. This reaction is called photosynthesis.

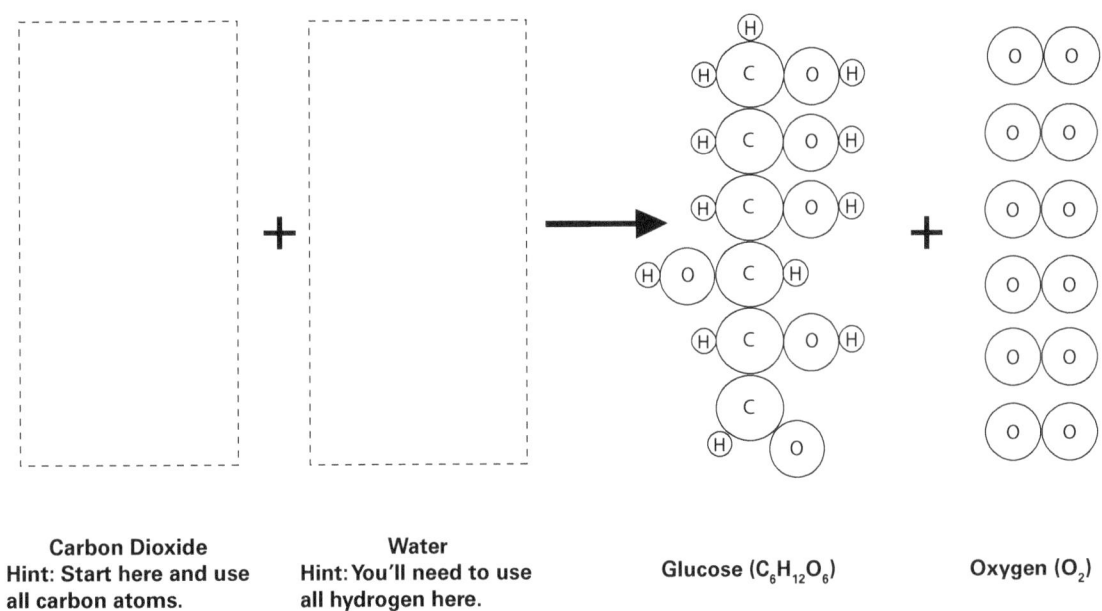

Carbon Dioxide
Hint: Start here and use all carbon atoms.

Water
Hint: You'll need to use all hydrogen here.

Glucose (C$_6$H$_{12}$O$_6$)

Oxygen (O$_2$)

6. Using your knowledge of the law of conservation of matter, balance this chemical reaction by drawing carbon dioxide and water molecules in the spaces provided.

Once Upon a Physical Science Book

Equation 4: Propane and Oxygen

Finally, propane can mix with oxygen to form carbon dioxide and water. Look at the following reaction and answer the questions that follow.

$$C_3H_8 + 5O_2 \longrightarrow 3CO_2 + 3H_2O$$

propane   oxygen   carbon dioxide   water

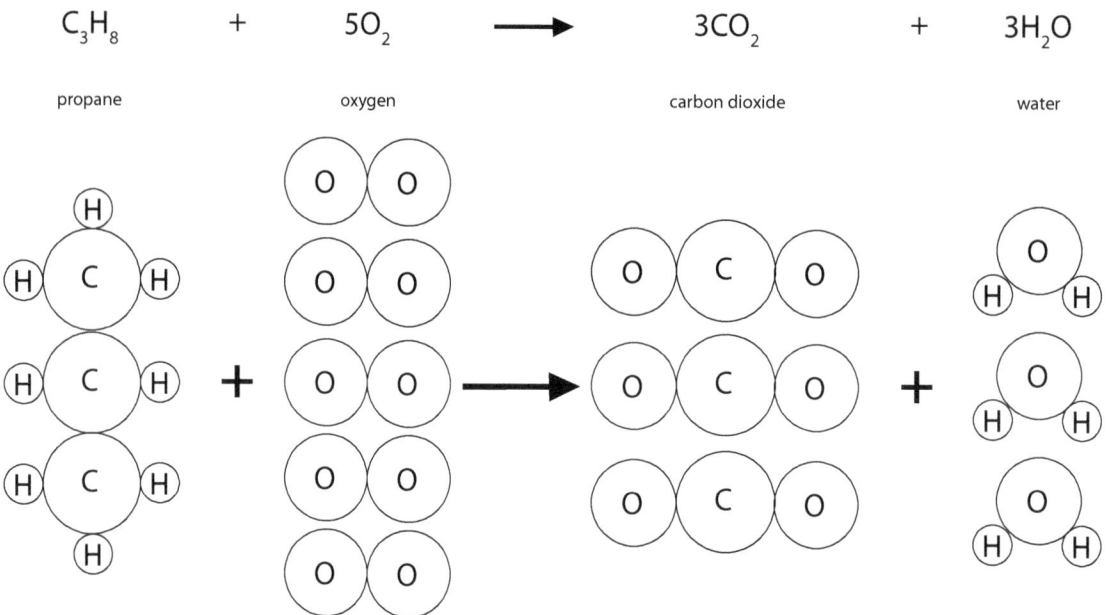

7. Is this chemical equation balanced?

8. How do you know?

Chapter 6
All Charged Up

Topics
- Static electricity
- Charges
- Electric fields

Reading Strategy
- Chunking

All Charged Up

Connections to Standards

Next Generation Science Standards (NGSS) Correlations

Standard
MS-PS2. Motion and Stability: Forces and Interactions (*www.nextgenscience.org/dci-arrangement/ms-ps2-motion-and-stability-forces-and-interactions*)

Performance Expectation(s)
The materials/lessons/activities outlined in this chapter are just one step toward reaching the performance expectations listed below.

MS-PS2-5. Conduct an investigation and evaluate the experimental design to provide evidence that fields exist between objects exerting forces on each other even though the objects are not in contact.

Dimension	Element	Matching Student Task or Question From the Activity
Science and Engineering Practice(s)	• Engaging in Argument From Evidence	• Students design a test using a balloon to determine the charge on a strip of tape. They use the evidence they generate to make an argument.
Disciplinary Core Idea(s)	**PS2.B.** Types of Interactions • Forces that act at a distance (electric, magnetic, and gravitational) can be explained by fields that extend through space and can be mapped by their effect on a test object (a charged object, a magnet, or a ball, respectively).	• Students observe the effects of static electricity as they look for evidence of an electric field around tape. • Students analyze a graph to show the relationship between forces exerted on two charged particles and the distance between them in the Thinking Mathematically section.
Crosscutting Concept(s)	• Cause and Effect	• Students use plastic tape to demonstrate the cause-and-effect relationships of variables that determine the strength of electric forces. • Students analyze data to provide evidence of an electric field (positive or negative) around plastic tape. • Students read about how the cause-and-effect relationships may be used to predict phenomena in natural or designed systems.

Common Core State Standards (CCSS) Correlations

Reading Standard(s)	• CCSS.ELA-Literacy.RST.6-8.9. Compare and contrast the information gained from experiments, simulations, video, or multimedia sources with that gained from reading a text on the same topic.	• Students use information from their lab and their reading to generate an argument about the charge on a strip of tape.
Writing Standard(s)	• CCSS.ELA-Literacy.WHST.6-8.1. Write arguments focused on discipline-specific content. • CCSS.ELA-Literacy.WHST.6-8.9. Draw evidence from informational texts to support analysis, reflection, and research.	• Students write arguments about whether tape and a paper strip have a field. • They also make an argument about the charge on a strip of tape. To support that argument, they must use information gleaned from the text.

Chapter 6

Background

Using static electricity to talk about fields can be a good entryway for an electromagnetism unit. Conveniently, rolls of clear plastic office tape demonstrate the phenomenon well. In this lesson, students will experiment with charges on tape. They will then read about how charged particles cause laundry mishaps. Finally, they will use what they have learned to determine the charge on a strip of tape and write an argument supporting their claim.

> **SAFETY NOTES**
> The following safety recommendations apply to all activities in this chapter:
> - Appropriately dispose of lab materials at the end of the activities as directed by the teacher.
> - Immediately report any lab accident to the teacher.
> - Wash your hands with soap and water after completing the activities.

Pre-Reading/Exploration

Materials for Activity (Per Group)

- 1 roll of clear office tape
- Scrap paper
- Access to desk or tabletop
- 1 ruler or meterstick

Activity

Use With Student Page(s): Stuck on Sticky Tape (lab sheet)

In this activity, students will explore the effect of static electricity and look for evidence of a field surrounding office tape. To engage their interest in the lab, begin by pulling a 10–12 cm strip of clear plastic tape from the roll. Show students how to fold the top a little, sticky side down, to make a nonsticky handle that won't cling to your fingers. Press the tape against a tabletop and then rip it off to create a tape strip with a strong charge. Show it to your students. Then, without using the term *static electricity*, carry the tape around the classroom and allow the nonsticky side to be attracted to various interesting objects (e.g., your face, a student's book bag, a window).

Ask students what they think is going on with the tape. Some students will suggest static electricity. Accept this proposal and any others as possibilities. Point out that the tape is moving even when it is not touching another object. Tell students that this type of behavior raised a question for scientists: How could an object exert a force on another object when they weren't even touching? Ask students if they have any suggestions for how an object can push or pull another without contact. Tells students they are going to explore some aspects of that question with today's lab.

Once Upon a Physical Science Book

Distribute the Stuck on Sticky Tape handout (p. 82) and have students complete the lab.

Reading

Use With Student Page(s): "One Clingy Sock" (article)

Introduce the Reading. Tell students that they will be reading an article that will provide clues about what was going on with the tape and why it behaved as it did.

Reading Strategy: Chunking

The strategy of chunking is introduced in Chapter 4. If you have not used Chapter 4 with your students, explain that many sentences in science writing are crammed with lots of ideas. It can be difficult to understand all the ideas in a sentence at one time. But if students break the sentence down into chunks, they can think about each piece individually. Model the technique for students. First, display the following sentence:

> *Objects with different charges, one positive and one negative, experience an attractive force.*

If you have introduced chunking before, ask a volunteer to add slash marks (/) in the places where he or she might chunk the sentence. Dissect the sentence for students, using the chunks that the volunteer selected. If you are introducing the strategy for the first time, place the slash marks yourself in the following locations:

> *Objects with different charges, / one positive and one negative, / experience an attractive force.*

Show that the phrase "one positive and one negative" is extra information placed in the middle of the sentence. Science writing frequently uses this kind of inserted phrase. Cover up the insertion, so the sentence reads:

> *Objects with different charges* ▬▬▬▬▬▬▬▬▬▬ *experience an attractive force.*

Read the sentence again to show that it makes sense without the inserted phrase. Then ask what information the inserted phrase adds to the sentence and ask why that information is important. Sentences with inserted phrases are called "interruption constructions," and they can be confusing. Students can always ignore the "interruption" and return to it after digesting the main sentence.

Chapter 6

Journal Question

After students have completed the reading, give them the following prompt for their reading journals, which will help them internalize the strategy they practiced: The interruption construction is a sentence in which extra information is inserted between commas. This construction can be useful in writing. You try it. Write a sentence that includes an interruption.

Application/Post-Reading/Writing

- **Writing Prompt.** Have students work with a rubber balloon and a roll of tape to respond to the following prompt: Rubber balloons pick up electrons very easily (and thus become negative). Inflate a balloon and rub it vigorously in your hair or on carpet. Make a fresh strip of ripped-from-the-desk tape (a.k.a. tape D) and use the balloon to figure out the charge on the tape. Make a claim and support it with evidence and reasoning. Include an explanation of what is happening in the molecules of the tape to create that charge.
    - **Pre-Writing Suggestions.** Have students spend a few moments thinking about what evidence they will need to gather to develop their responses, and allow them time to conduct their tests. Ask students to list science concepts they might need to explain in their responses. Then have them number the concepts in the order they want them to appear. Tell the students that the concepts should be presented in an order that makes the responses as clear and logical as possible.
    - **Key Evaluation Point.** The charge on tape D may vary depending on the tape you use and the table surface, so accept any claim that is supported by evidence. Because opposite charges repel and like charges attract, students should claim that the tape is negative if it is repelled by the balloon and positive if it is attractive. In their responses, students should include mention of either the loss or gain of electrons on tape D to explain the creation of the charge.
- **Thinking Mathematically.** Students will examine a graph in the Electric Field Forces worksheet (p. 88) to see how electric fields change over distance.

All Charged Up

Stuck on Sticky Tape

1. Start exploring.

    a. Get a piece of tape that is approximately 8–10 cm long. Fold one end over, sticky side down, to make a nonsticky "handle" that won't cling to your fingers.

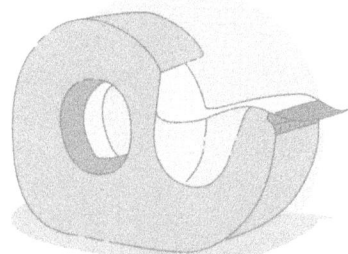

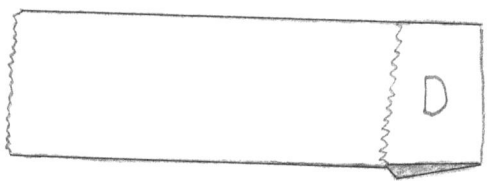

    b. Put the tape, sticky side down, on a clean, dry desk or table. Label the tape "D" for "desk."

    c. Rip the tape off the table. Then stick it to the edge of the table so that it can hang freely.

    Choose some objects, such as a pencil, a cup, your finger, or whatever else you have available. Slowly bring each one close to—but not touching—the tape. Describe what happens to the tape.

2. For some reason, the tape is interacting with the objects even when they aren't touching. Figure out if this same reaction happens with paper. Cut out or tear off a piece of scrap paper that is about the same size as the tape strip. Use a small bit of tape to hold your paper strip to the desk so it can hang freely. Bring the same objects you used in Part 1 close to the strip of paper. Describe what happens to the paper. Does it behave the same way as the strip of tape?

Chapter 6

3. If an object can exert force on another object *that it is not touching*, it is said to be surrounded by a "field." Does tape D generate a field? _____ What is your evidence?

   Does the strip of paper generate a field? _____ What is your evidence?

4. Using your ruler, figure out a way to explore the size of the field around tape D. (Feel free to make a fresh tape D if needed.) How close must the tape and an object be to interact? What is your evidence?

   Does it matter if the tape was just pulled from the table or if you have been using it for other tests? What is your evidence?

5. Pull off a new piece of tape and make a handle. Put the tape on the table and label it "D2". Rip it off. Slowly bring the nonsticky side of tape D2 near the nonsticky side of tape D. (Use the nonsticky sides so they don't get stuck together). What happens?

6. Now explore whether all tape strips generate the same kind of fields.
   - Pull off a new piece of tape, make a handle, and stick it to the table. Label it "B" for "bottom."
   - Pull off another piece of tape, make a handle, and put it directly on top of tape B, sticky side down. Label it "T" for "top."
   - Gently pull the B-T tape combination off the table.
   - Touch the sticky side of the tape combination with your hand, taking care not to wrinkle the tapes. Stick the B-T tape combination to the edge of the table.

Once Upon a Physical Science Book

All Charged Up

a. How can you test if the B-T tape combination has a field?

b. Do your test. Does the B-T tape combination have a field? What is your evidence?

7. If the B-T tape combination has a field, then touch it several times with your finger and test it again until it is no longer exhibiting a field. Now vigorously rip tape B and tape T apart and hang the tapes from the edge of the table. Prepare a fresh tape D as well.

   a. Describe the interactions when the following pairs are brought together:
      * Tape B with a fresh tape D

      * Tape T with a fresh tape D

      * Tape B and tape T

   b. Do tape B and tape T generate the same kind of field? What is your evidence?

8. The field around plastic tape is created by static electricity. Electric fields can have a positive or a negative charge. When two matching fields interact, the objects repel each other. Which of the three tapes (B, T, and D) have the same charge? _____ What is your evidence?

Chapter 6

One Clingy Sock

When Gabriela came to breakfast, her little sister Maya began to giggle. "I usually wear my socks on my feet, but you've got one on your back," Maya laughed. Gabriela tugged her fleece hoodie around to get a look. Sure enough, a small pink sock was clinging to the back of it. She grabbed it, and it came off easily. "Static," she said as she chucked the sock at her still-giggling sister.

> **REMEMBER YOUR CODES**
> ! This is important.
> ✓ I knew that.
> X This is different from what I thought.
> ? I don't understand.

You may have had a similar experience with clothes sticking together. Or you may have felt a shock while sorting clothes fresh from the dryer. You may have even seen a small item, such as a sock "hop" onto a larger item, such as a blanket. Like Gabriela, you probably know you can chalk these strange experiences up to static electricity. But what is static electricity, and how can it cause a sock to link up with a hoodie?

An Atomic Problem

Static electricity begins in the tiny atoms that make up the fibers woven into your clothes. Remember that atoms are made of three particles: protons, neutrons, and electrons. Protons have a positive charge. Neutrons are neutral, meaning they have no charge. Electrons have a negative charge. The number of protons and electrons in an atom are equal—so taken together, the atom is neutral.

Figure S6.1. Atom

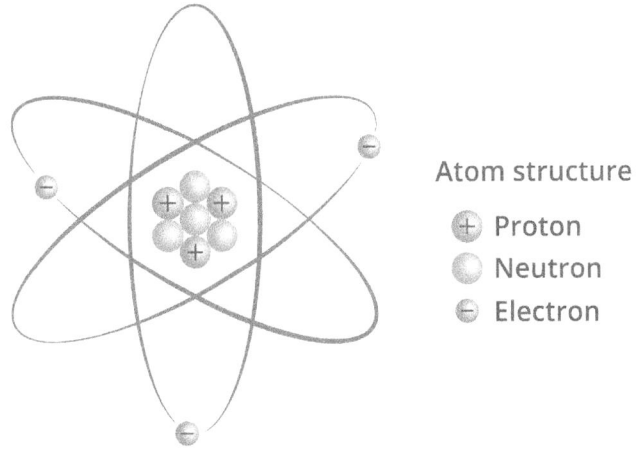

Atom structure
⊕ Proton
○ Neutron
⊖ Electron

Negatively charged electrons are outside the nucleus in an atom.

However, while the protons and neutrons are crammed into the nucleus together, electrons travel around the outside of the atom (Figure S6.1). This distance means that electrons sometimes separate from their atoms.

Take the atoms in Gabriela's sister's sock, for example. As it tumbled in the dryer, the sock rubbed against other types of clothes made of different types of molecules. The atoms in some molecules give away their electrons easily. The atoms in other molecules grab extra electrons.

Once Upon a Physical Science Book

85

All Charged Up

Suppose the sock is made of a material that gives away electrons, whereas Gabriela's hoodie is made of a material that grabs electrons. When they rubbed together in the dryer, the sock gave electrons to the hoodie. As a result, the hoodie had extra negatively charged electrons, which gave the fabric a negative charge. The sock, on the other hand, was left with more protons than electrons. Therefore, the sock had a positive charge. (See Figure S6.2.)

Electric Fields

Atoms and molecules with a charge generate an electric field. Within that field, they can interact with other objects, even if they aren't touching. Objects with different charges, one positive and one negative, experience an attractive force (Figure S6.3). Objects with the same charge, both positive or both negative, repel each other. Since Gabriela's hoodie had a negative charge and the sock had a positive charge, they stuck together.

Figure S6.2. Static Buildup

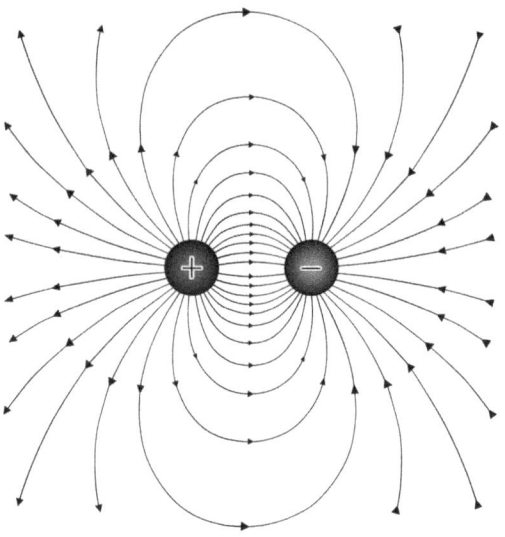

Figure S6.3. Electric Field of Two Unlike Charges

Staticky Sticky Tape

Later in science class, Gabriela needed to tape some lab notes into her science notebook. She pulled a strip of clear plastic tape from the roll. As she pulled it off, the strip of tape lost some electrons to the layer below and became positively charged. Before she could smooth it onto her paper, the strip folded itself backward and stuck to her hand.

While this was aggravating for Gabriela, it demonstrated another property of electric charges: An object with an electric charge can interact with an object that is neutral, like Gabriella's skin. The positively charged tape was attracted to the electrons in the atoms of Gabriella's skin, and the force of the electric field was enough to pull the tape toward her hand.

Chapter 6

Electrons on the Move

When Gabriela got home from school, she generated one more electric field—she turned on her lights. In this case, however, she wasn't dealing with static electricity. Static means "not moving." To power her lights, the electrons were moving, flowing through the wires. Moving electrons create the controlled electricity that we use every day. But Gabriela wasn't too worried about electricity, static or otherwise. She just wanted to make sure the laundry was folded and put away, with no more clingy socks.

The Big Question

Gabriela's little sister wants to know why her pink sock stuck to Gabriela's hoodie. How would you explain it to her?

Thinking Mathematically: Electric Field Forces

You saw in the lab that two charged pieces of tape interact but only if they are close together. The graph in Figure S6.4 shows the force that two charged particles exert on each other at various distances.

Figure S6.4. The Relationship Between Force and Distance Between Charges

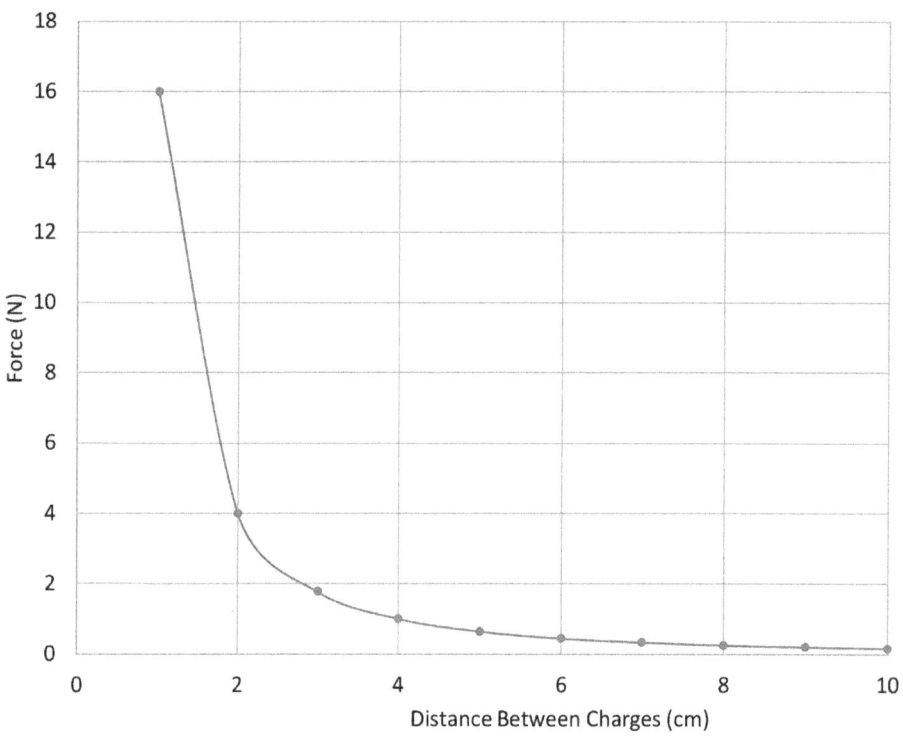

1. Describe the shape of the graph.

2. As the charges get closer together, what happens to the force of their attraction?

Chapter 6

3. Imagine that this chart represents the force between two charged pieces of tape. If you hold the tape strips 10 cm apart and move them to 8 cm apart, how does the force change?

4. Now imagine that you are holding them 4 cm apart and move them to 2 cm apart. How does the force change?

5. Think about your experiences with gradually moving charged tape pieces closer and closer together. How does this graph relate to what you observed with the tape?

Chapter 7

Inertia: It's a Drag

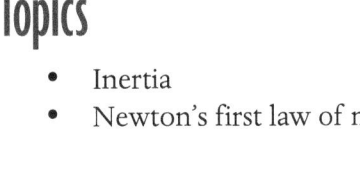

Topics
- Inertia
- Newton's first law of motion

Reading Strategy
- Previewing diagrams and illustrations

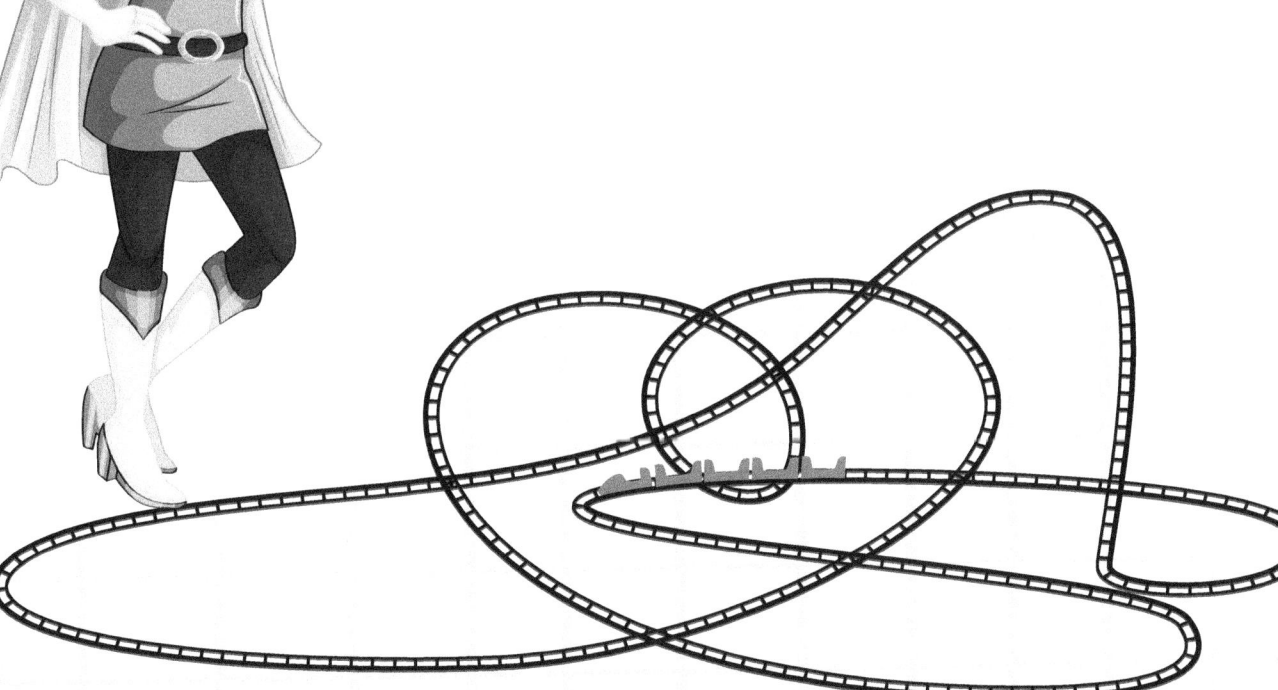

Inertia: It's a Drag

Connections to Standards

Next Generation Science Standards (NGSS) Correlations	
Standard **MS-PS2.** Motion and Stability: Forces and Interactions (*www.nextgenscience.org/dci-arrangement/ms-ps2-motion-and-stability-forces-and-interactions*)	
Performance Expectation(s) The materials/lessons/activities outlined in this chapter are just one step toward reaching the performance expectation(s) listed below. **MS-PS2-2.** Plan an investigation to provide evidence that the change in an object's motion depends on the sum of the forces on the object and the mass of the object.	

Dimension	Element	Matching Student Task or Question From the Activity
Science and Engineering Practice(s)	• Using Mathematics and Computational Thinking	• Students analyze force diagrams in the Thinking Mathematically section to determine the motion of an object when the forces in a system are balanced or unbalanced.
Disciplinary Core Idea(s)	PS2.A. Forces and Motion • The motion of an object is determined by the sum of the forces acting on it; if the total force on the object is not zero, its motion will change. The greater the mass of the object, the greater the force needed to achieve the same change in motion. For any given object, a larger force causes a larger change in motion. • All positions of objects and the directions of forces and motions must be described in an arbitrarily chosen reference frame and arbitrarily chosen units of size. In order to share information with other people, these choices must also be shared.	• Students observe the change in motion of a ball on a piece of construction paper and make connections to traveling in a car. • Students observe the movement of marbles (with different masses) through a maze after applying a force on them by blowing through a straw. • Students read about what happens when you ride on a school bus and how inertia affects the motion of objects like marbles.
Crosscutting Concept(s)	• Stability and Change	• Students analyze the motion of a ball to observe how inertia affects its motion. • Students analyze data to determine how the force applied to marbles of varying mass affect the change in motion and direction of the marbles. • Students analyze force diagrams to determine when the forces in a system are balanced or unbalanced.
Common Core State Standards (CCSS) Correlations		
Reading Standard(s)	• CCSS.ELA-Literacy.RST.6-8.7. Integrate quantitative or technical information expressed in words in a text with a version of that information expressed visually (e.g., in a flowchart, diagram, model, graph, or table).	• The reading strategy for this chapter is previewing diagrams and illustrations. • The journal question asks students to compare what they learned from a diagram to what they understood from the text.

Continued

Chapter 7

Writing Standard(s)	• CCSS.ELA-Literacy.WHST.6-8.2. Write informative/explanatory texts, including the narration of historical events, scientific procedures/experiments, or technical processes.	• Writing prompt: When the ketchup level in a bottle gets low, people will often solve the problem by turning the bottle upside down, giving it a hard shake, and then stopping the bottle suddenly. Explain why this can get the ketchup to the end of the bottle.

Background

This chapter focuses on helping students develop an intuitive understanding of inertia and the role it plays in motion. You can use it near the beginning of a unit on force and motion or after you have already talked about describing motion mathematically and introduced the concepts of velocity and acceleration. Students will experiment with the motion of a ball and marble, read about how similar forces are at work in a vehicle, and write about how an understanding of forces can help rescue the last bits of ketchup from a bottle.

> **SAFETY NOTES**
> The following safety recommendations apply to all activities in this chapter:
> - Wear safety glasses with side shields or safety goggles during the setup, hands-on, and takedown segments of the activities.
> - Pick up any items that have dropped on the floor to avoid tripping and falling hazards.
> - Appropriately dispose of lab materials at the end of the activities as directed by the teacher.
> - Immediately report any lab accident to the teacher.
> - Wash your hands with soap and water after completing the activities.

Pre-Reading/Exploration

Materials for Activity (Per Group)

- Safety glasses with side shields or safety goggles
- 1 large (11" × 17") sheet of foam board (best option), cardboard, or card stock
- 1 basketball, volleyball, playground ball, or something similar
- 1 roll of masking tape
- Tabletop or hard floor
- Dry erase markers or chalk
- 1 ruler or meterstick
- 1 standard marble
- 1 large shooter marble
- Straws for each person
- 1 stopwatch

Activity

Use With Student Page(s): Hold On! (lab sheet for Part 1) and Marble Maze (lab sheet for Part 2)

This exploration of Newton's first law and inertia has two parts. In the first, students will experiment with pulling a ball around on a piece of

Inertia: It's a Drag

foam board. In the second, students will try to send a small and a large marble through a maze.

Part 1: Hold On!

In Step 1 of this activity, students will try to keep a ball balanced on a sheet of foam board, cardboard, or card stock as they pull the board in a straight line. Step 2 challenges them to turn the board as quickly as possible without losing the ball. In Step 3, students will observe that the ball rolls on if they suddenly stop moving the board.

After students have completed their explorations, talk with them about their experiences riding in a car or bus. Connect these prior experiences to what they observed during the lab. (Those who have ridden a bus while holding onto a bar or grab handles will have particular insight into how bodies respond to vehicle movement.) To solidify their understanding, ask students the following questions, which correspond to different steps of the activity:

- **Question 1 (Similar to Step 1): How do you move when a car accelerates?** Students will likely describe feeling as if they are being pushed back into their seats. Ask them to consider the ball simulation. See if they can figure out from the simulation that the seat of a car presses into a rider, pushing him or her to keep up with the car.
- **Question 2 (Similar to Step 2): How does your body move when the car takes a sharp turn?** Students may say that they are thrown to the side of the car. Have them revisit Step 2 to see if they can determine that their bodies are continuing in a straight-line path but "run into" the door that has found its way in front of them as the car turns.
- **Question 3 (Similar to Step 3): How do you move when the car makes a sudden stop?** Students will answer that they feel as if they are thrown forward. Have them use the simulation to figure out that, in reality, they are already moving forward when the car stops. And instead of immediately stopping with the rest of the car, they continue to move forward.
- **Question 4 (Similar to Step 4): What helps keep your motion in line with the car's motion?** Just as setting the ball in the ring of tape increases friction and helps keep the ball on the paper, friction between the students' bodies and their car seats does a lot to keep them moving with the car. However, very aggressive changes will still cause the passenger (or, in the case of the simulation, the ball) to move independently. Most students will create a tape "seatbelt" in the final challenge to increase the ability of the ball to stay in

its seat during intense changes, just as a seatbelt in a car protects people from flying through the windshield after a sudden stop.

Part 2: Marble Maze

For this activity, students will need to draw mazes for their marbles. See Figure 7.1. Dry erase markers can be easily wiped off most science lab tables, but test yours in an inconspicuous spot to be certain. A marble rolling on a lab tabletop experiences very little friction, and a dry erase marker creates a completely smooth border, so the combination works well. However, if you do not have lab tables, or if the dry erase markers don't erase well from your tables, you can make substitutions. Try chalk instead of dry erase marker or a large sheet of newsprint instead of a naked tabletop. The friction in these setups will be higher. Try to create the setup with the least friction possible for your setting.

Figure 7.1. Marble Maze on a Science Lab Table

With their mazes, students will be exploring acceleration in all its forms. They will also be getting a visceral feel for how changing the mass of the marble affects its inertia; namely, they will have to blow much harder through their straws when they switch to the larger marble.

Once your students have had some success but could use a break, ask, "When does the marble roll become a challenge?" (Students may say it becomes a challenge when the marble has to turn a corner, when they have to slow the marble down, etc.) Then ask, "How has your strategy changed as you have practiced?" (Strategy changes may include blowing through the straw more gently, taking more time on the turns, slowing down before the turns, etc.) When students are ready, encourage them to move to the bigger marble and then trade tables with another group.

Conclude the exploration by asking, "What was different about using the larger marble?" (Students may say that it was harder to get moving, harder to turn, required more blowing, etc.) Then ask, "What did you notice about the big marble as you a made turns?" (Answers may include that the marble couldn't go as fast, was harder to keep in the maze, etc.)

Reading

Use With Student Page(s): "One Long Bus Ride" (article)

Introduce the Reading. Tell students that they are going to read more about how motion works. In particular, this article will talk about what happens as you ride in a vehicle. Remind students that they can picture the ball sliding on the foam board if they need a mental image of what happens to a passenger in a car or bus.

Reading Strategy: Previewing Diagrams and Illustrations

Tell students that in some books, they can just glance at the pictures and keep on reading. But in science text, pictures and diagrams often carry more information than what is written or show the information in a way that words can't capture. When reading science text, it is important not to skip over the images. In fact, studying some of the images ahead of reading can make the text easier to understand. As they read this article (and other articles), they should take a moment to study each picture. If the text refers to a picture, it is a cue to stop, look, and analyze what the text is talking about. With complicated systems, it can be easier to make a mental image of a diagram than it is to remember written descriptions.

Have students sit in pairs for this discussion. One student should be partner A and the other should be partner B. Lead the pairs through the following conversation, allowing one pair to share after each question.

Begin by displaying Figure S7.2 from "One Long Bus Ride" (p. 100) *without* its title.

- Have the As describe to their partners what they see in this diagram. If they don't know the words for certain elements or what some parts of the diagram might represent, that's OK. Their main goal is to make observations.
- Have the Bs make a prediction: What do they think the diagram has to do with the lab they just completed? Once again, they don't need to be correct. Affirm any reasonable response, but say they will soon have a chance to read an article that will give them more answers.

Now display Figure S7.3 (p. 101) but cover up the label "friction with the table."

- This time, have the Bs describe what they see in the diagram.

- Have the As predict what the covered-up label might say. Once again, don't confirm or deny possibilities. Let students know that they will be able to read the label in the article.

Display Figure S7.4 (p. 101) *without* its title.

- Ask the Bs to describe what they see to their partner.
- Have the As make a prediction: What idea do you think this diagram is trying to show?

Give students the article "One Long Bus Ride" and tell them to check how accurate their predictions were as they read.

Journal Question

After students have completed the reading, give them the following question for their reading journals, which will help them internalize the strategy they practiced: Before reading, you previewed the diagrams. Pick one of the diagrams that you previewed. Did you find the diagram easier to understand or the text? Why?

Application/Post-Reading/Writing

- **Writing Prompt.** When the ketchup level in a bottle gets low, people will often solve the problem by turning the bottle upside down, giving it a hard shake, and then stopping the bottle suddenly. Explain why this can get the ketchup to the end of the bottle.
  - **Pre-Writing Suggestions.** Do a quick demonstration with a bottle of ketchup to illustrate the actions described in the prompt. Suggest that students think about the ball and paper simulation as they work out their answer. Ask, "Which step in the ball activity was most like this situation?" Also have students consider what science words they should include in their answers and how a diagram could improve their explanation.
  - **Key Evaluation Point.** Students should say that the ketchup has inertia. Moving the bottle puts the ketchup into motion. When the bottle stops suddenly, the ketchup's inertia keeps it moving so it sloshes to the spout end of the bottle.
- **Thinking Mathematically.** Students will use the Out of Balance? worksheet (p. 103) to examine the effect of different forces on an object.

Inertia: It's a Drag

Hold On!

Step 1: Place the ball on the middle of your foam board, cardboard, or cardstock. (This will be called the platform.) Pull the platform as quickly as you can in a straight line while keeping the ball in the middle of the platform. See how quickly you can pull it to the end of your table. What happened to the ball if you pulled the platform too fast?

Step 2: Get the ball and platform moving again in a straight line. Change the direction that the platform is moving in as quickly as you can without losing the ball. You can choose to turn right or left. What did the ball do if the turn was too fast?

Step 3: Get the platform and ball moving as fast as you can and stop it as quickly as you can. What did the ball do if you stopped the platform too quickly?

Step 4: Place the roll of tape flat on its side in the middle of the paper. Attempt all the moves from Steps 1, 2, and 3 with the ball resting in the center of the tape roll.

- What effect did the tape roll have on how quickly you were able to make changes (speed up, turn, and slow down)?

- What else could you do to help the ball stay on the platform during sudden changes in motion? Sketch your solution on the back of this page.

- Try your solution. How did it work?

Chapter 7

Marble Maze

1. Using the chalk or dry erase marker your teacher supplied, draw a maze on your table that fits the following requirements:

   - There are walls on both sides of the path.
   - The path is 8–10 cm wide.
   - There are between three and five 90-degree turns (see Figure S7.1).

2. Use your straws to blow the small marble through your maze. If you cross a line, you must return to the start. You can work as a team or individually. Practice until you can get the marble all the way through.

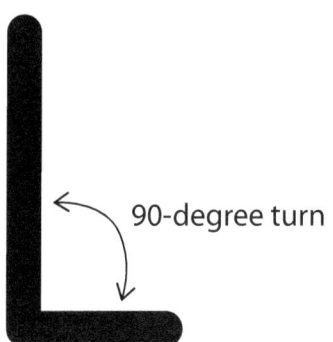

Figure S7.1. 90-Degree Turn

3. Time how quickly you can complete the maze. Use the dry erase marker or chalk to write your best time on your table next to your maze.

4. Switch to the larger marble and repeat Steps 2 and 3. How was the experience different when you used the larger marble?

5. If you have time, trade mazes with another group and see if you can beat their best times with the small and large marble.

Inertia: It's a Drag

One Long Bus Ride

Field trip to the science museum! You load onto the school bus, packed three to a seat. You find yourself wedged between Football Fred, the biggest student in your class, and Tiny Tiana, the smallest. Ms. Wheeler is driving. You groan. Ms. Wheeler has only been driving buses for a few weeks, and she's not exactly smooth at the wheel.

> **REMEMBER YOUR CODES**
> ! This is important.
> ✓ I knew that.
> X This is different from what I thought.
> ? I don't understand.

"Hang on!" she cries as she hits the gas. The bus leaps out of its parking spot. As the bus plows forward, you lean back hard, feeling as though you were being pressed to the back of your seat. The bus races for the end of the parking lot, then—*screech!*—Ms. Wheeler hits the brakes. You barely stop yourself from banging your face on the back of the seat in front of you.

Things Keep Doing What They Are Doing

Why all the lurching back and forth? You can thank inertia. Inertia describes how objects resist changing speed or direction. Before the bus started, you were sitting still in your seat. The bus moved forward, but your body resisted getting moving. That's inertia. The back of your seat had to push you forward to help you get with the program, as seen in Figure S7.2. Once your body was moving along with the bus, it was ready to continue moving forward. When the bus stopped suddenly, your body kept going, propelling your face toward the seat in front of you. That's also inertia.

Figure S7.2. Inertia as the Bus Speeds Up

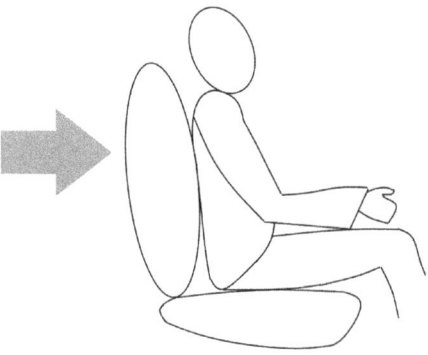

The bus pushes against you as it starts forward.

Isaac Newton, an early physicist, described inertia this way in his first law of motion: An object in motion remains in motion, and an object at rest remains at rest, unless acted upon by an external force. Imagine that you set a marble on a smooth table. As long as you don't touch it or blow on it or tilt the table, it is going to sit in that same spot. It is "at rest," and it is going to stay that way. But say you take a deep breath and blow on the marble. Now you have applied an outside force that is strong enough to overcome the marble's inertia. The marble starts rolling forward, and it will keep rolling forward, even if you don't puff on it again. This time, its inertia keeps it moving.

If you have a long enough table, the marble will eventually stop—not because of some failure of inertia, but because there is a small force still acting on the marble. That force is friction, as shown in Figure S7.3. Friction between the marble and the table gradually slows the marble.

Figure S7.3. Friction Gradually Slows the Marble

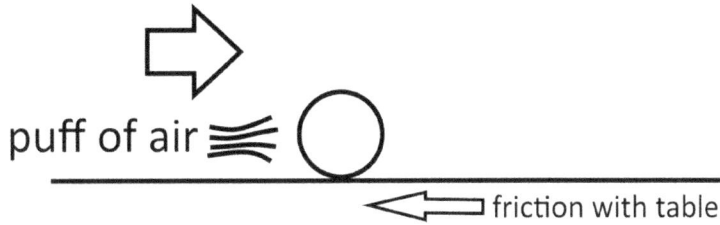

Taking Inertia for a Spin

Back on the bus, you're bracing yourself to try to limit the effects of inertia. "Oh! I almost missed my turn," shouts Ms. Wheeler, as she wrenches the bus to the right. You find yourself leaning left against Football Fred, who leans against the window. At the same time, Tiny Tiana falls over onto you. Why, when the bus turned *right*, did you and your seatmates fall *left*?

Once again, the answer is inertia. See Figure S7.4. Before the turn, you and Tiana were moving steadily forward. When the bus turned, you kept moving in that same direction. But because of the turn, Football Fred was sitting where "forward" used to be!

Figure S7.4. Inertia as the Bus Turns

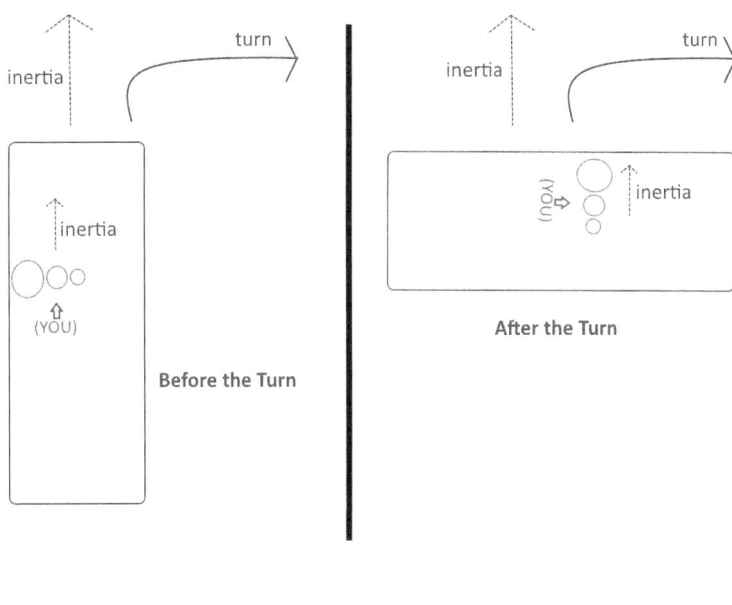

Inertia: It's a Drag

Big Guys Have Big Inertia

You recover your balance and try to act relaxed, like you weren't just having a lean-fest with Tiana and Fred. But then, Ms. Wheeler pulls another wild turn, this time to the left. Everyone in your seat flies right, and Football Fred slams into you. Whoa! It was awkward when Tiny Tiana was snuggling your shoulder, but with Football Fred, you feel like you've been hit by a boulder.

What caused this unintentional tackle? You guessed it—inertia! The more mass an object has, the greater its inertia. Football Fred has more mass than Tiny Tiana, so Football Fred also has more inertia. It is going to take an even greater outside force to get Fred to turn with the bus. And where is he getting that outside force? From your poor, bruised shoulder.

Imagine again a marble rolling on a table. It only takes a small puff of air to get a small marble rolling. However, if you have a marble with more mass, it's going to take a stronger puff of air. Once the more massive marble is moving, it's also going to take more force to stop it.

In fact, mass and inertia are essentially two ways of describing the same property. If you know how massive an object is, you also know something about how hard the object is going to be to move. To measure an object's inertia, you simply report its mass.

Even Buses Have Mass

Unfortunately for Ms. Wheeler, a school bus full of students has a lot of mass. It takes a lot of force from the brakes to stop it—force that Ms. Wheeler does not apply quite soon enough as you arrive at the science museum. The bus jumps the low curb, trundles over a cluster of daffodils and two azalea bushes, and shudders to a stop.

As you get off the bus, you feel like a marble exiting a pinball machine. You don't need to go to the science museum; you've had your science lesson on the ride over. On the other hand, maybe the museum is a good idea. You don't really want to start the ride home just yet.

The Big Question

How does inertia affect the motion of a passenger in a school bus?

Chapter 7

Thinking Mathematically: Out of Balance?

A resting object will keep resting and a moving object will keep moving unless it encounters an unbalanced force. Look at the boxes pictured in the card set below and observe the forces they are experiencing. (Note that forces are measured in Newtons [N].) How will each box move—to the right, to the left, or will it stay in place? Cut out the cards and sort them into the three categories listed. Then circle the category that shows balanced forces.

Box Will Move Right	Box Will Move Left	Box Will Stay in Place

Forces on Boxes Card Set

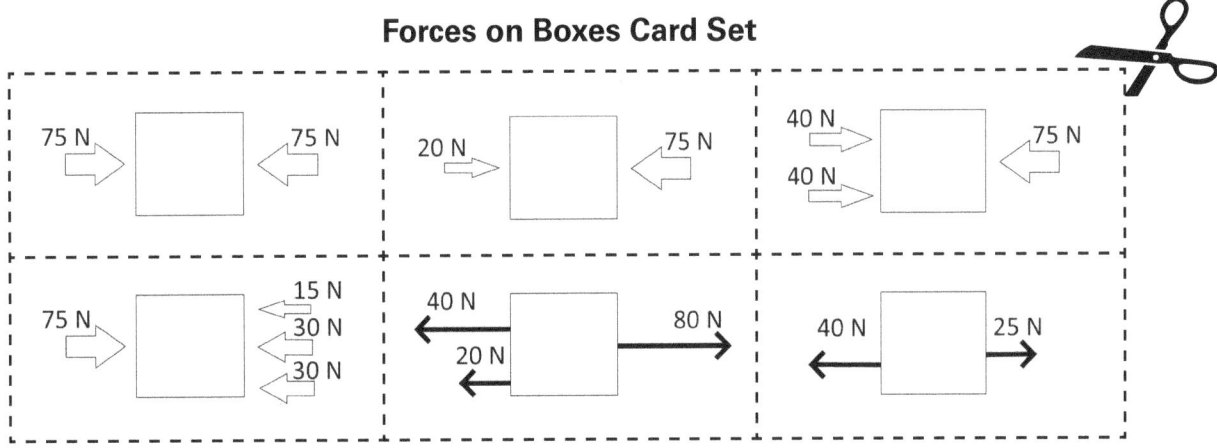

Once Upon a Physical Science Book

103

Kick Force

Chapter 8

Topics
- Newton's second law of motion
- Animal kicks

Reading Strategy
- Reading technical text

Kick Force

Connections to Standards

	Next Generation Science Standards (NGSS) Correlations	
Standard		
MS-PS2. Motion and Stability: Forces and Interactions (www.nextgenscience.org/dci-arrangement/ms-ps2-motion-and-stability-forces-and-interactions)		
Performance Expectation(s)		
The materials/lessons/activities outlined in this chapter are just one step toward reaching the performance expectation(s) listed below. **MS-PS2-2.** Plan an investigation to provide evidence that the change in an object's motion depends on the sum of the forces on the object and the mass of the object.		
Dimension	**Element**	**Matching Student Task or Question From the Activity**
Science and Engineering Practice(s)	• Planning and Carrying Out Investigations • Engaging in Argument From Evidence	• Students plan and conduct an investigation to test the effects of force on the distance traveled by a plastic cap. • Students make an argument about the optimal setup (force) for shooting a puck a specific distance.
Disciplinary Core Idea(s)	**PS2.A.** Forces and Motion • The motion of an object is determined by the sum of the forces acting on it; if the total force on the object is not zero, its motion will change. The greater the mass of the object, the greater the force needed to achieve the same change in motion. For any given object, a larger force causes a larger change in motion.	• Students conduct an investigation to measure the effects of force on the distance traveled by a plastic cap. • Students observe the effects of equal force on objects of different masses (i.e., a golf ball and a Ping-Pong ball). • Students read about how to apply Newton's second law to calculate the force, mass, or acceleration of an object in motion.
Crosscutting Concept(s)	• Cause and Effect	• Student use data from a graph to examine how force applied to a puck affects the distance the object travels.
	Common Core State Standards (CCSS) Correlations	
Reading Standard(s)	• CCSS.ELA-Literacy.RST.6-8.4. Determine the meaning of symbols, key terms, and other domain-specific words and phrases as they are used in a specific scientific or technical context relevant to grades 6–8 texts and topics. • CCSS.ELA-Literacy.RST.6-8.7. Integrate quantitative or technical information expressed in words in a text with a version of that information expressed visually (e.g., in a flowchart, diagram, model, graph, or table).	• The reading skill for this chapter is integrating information from text and mathematical representations in the form of symbols, equations, and sample problems.
Writing Standard(s)	• CCSS.ELA-Literacy.WHST.6-8.1. Write arguments focused on discipline-specific content.	• Students make an argument about the effect of force and mass on the distance traveled by a plastic cap.

Chapter 8

Background

This chapter focuses on the mathematical understanding of $F = m \times a$. Students will conduct an experiment to look at the relationship between a plastic cap's force, mass, and movement across a table. Then they will read about researchers using Newton's second law to understand animal motion and practice using the equation $F = m \times a$.

Pre-Reading/Exploration

Materials for Activity (Per Group)

- Safety glasses with side shields or safety goggles
- 1 golf ball
- 1 Ping-Pong ball
- 1 rubber band launcher (See Figure 8.1.)
- 1 cap from a sports drink bottle or similar beverage container (roughly 4 cm in diameter)
- 10 g of modeling clay
- Flat surface such as a lab table
- 1 cleanable writing utensil (wet erase marker, dry erase marker, or chalk; test selection on the flat surface prior to activity)
- 1 meterstick

> **SAFETY NOTES**
> The following safety recommendations apply to all activities in this chapter:
> - Wear safety glasses with side shields or safety goggles during the setup, hands-on, and takedown segments of the activities.
> - Make sure the trajectory for the projectile is marked off and do not allow anyone to stand in its path.
> - Appropriately dispose of lab materials at the end of the activities as directed by the teacher.
> - Immediately report any lab accident to the teacher.
> - Wash your hands with soap and water after completing the activities.

Activity

Use With Student Page(s): Shuffleboard Hero (lab sheet)

This exploration asks students to experiment with changing force and changing mass on a moving bottle cap. There are two important "thinking parts" to this lab: designing the experiment and analyzing the data. Actual data collection can become quite tedious and take time that would be better spent on the thinking parts. Therefore, we suggest having students think through the experimental design but then analyze the sample data provided in Figure S8.2 (p. 113). You can, of course, have students analyze and collect their own data.

Before class, make a set of rubber band launchers by wrapping one rubber band around another, as shown in Figure 8.1. To use the launcher, hold the two ends of

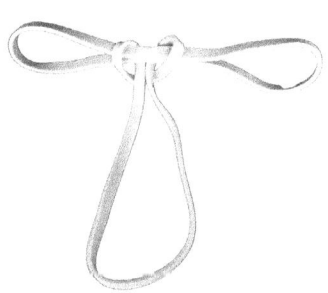

Figure 8.1. How to Make the Launcher

Kick Force

Figure 8.2. Holding the Launcher

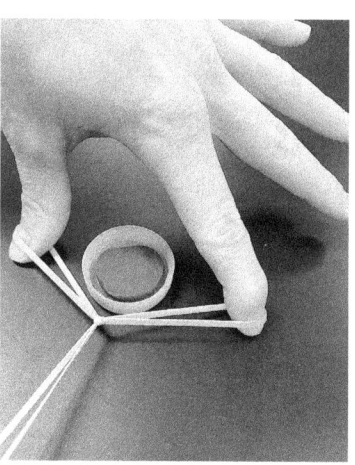

Figure 8.3. Launcher With Balls

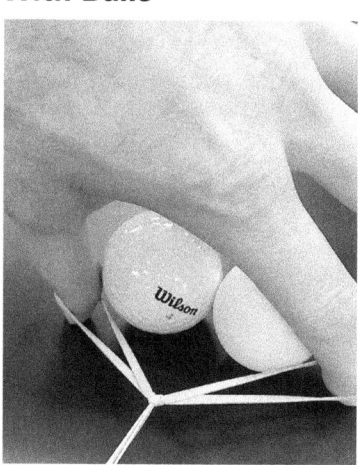

the same rubber band between your index finger and thumb (Figure 8.2). Place the rubber band directly behind the two balls. Pull back the attached rubber band and release. Practice using the launcher before class until you can shoot the golf ball and the Ping-Pong ball simultaneously (Figure 8.3).

Introduce the lesson by shooting the golf ball and Ping-Pong ball at the same time. Help students observe that the force on the two balls was the same (i.e., both received a push from the same launcher at the same time). Ask, "Why did the two balls have such different flights?" Note that students might use the word *weight* in their answer; point out that the term they should be using here is *mass*.

Show a short video clip of a shuffleboard game and point out that if you could figure out how to always land the puck in the right place, you'd win. Give each group a rubber band launcher, a plastic cap (their puck), modeling clay, a marker, a meterstick, and goggles for each group member. Tell students to try to figure out how to get their puck to slide exactly 40 cm down their lab tabletops. Allow 5–8 minutes for students to experiment and then call them back together.

Ask, "What variables could you change to get the puck to move more or less?" (Answers: They could change the force by varying how far back they pull the launcher; they could also change the mass by adding clay to or subtracting clay from the puck.) Then ask, "What difficulties did you have in hitting the 40 cm mark?" (Answers: Students may say that it was hard to use the launcher consistently, hard to find a way to measure how far back they were pulling the rubber band, etc.) Finally, ask, "How could you measure the differences in the amount of force you applied to the puck?" (Students may have a variety of answers; accept any that make sense.)

> **SAFETY NOTE**
> Make sure there are no students or breakable objects in the direction that you shoot the balls.

> **SAFETY NOTE**
> Require all students to wear goggles while **any** group is still experimenting, and have them make sure no one is in the way each time they launch.

Tell students that you want to calculate the best combination of force and mass to use to have the best chance of hitting 40 cm. To do that, you'll need to design an experiment showing the ways force and mass affect how far the puck travels. Have students work with you to come up with a possible experimental design. The exact design can vary, but it should include some way to measure the force of the launcher (perhaps by measuring how far you pull the launcher back). It should also include launching with the same force but changing the mass between trials and then changing the mass while holding the force constant. Students will want to plan for several trials of each.

Reading

Use With Student Page(s): "Of Kangaroos and Secretary Birds" (article)

Introduce the Reading. Tell students they are going to read about how scientists use the relationship between force, mass, and acceleration in research on animal motion.

Reading Strategy: Reading Technical Text

Reading texts with math problems requires a different approach from reading pure text. If you talked with your students about using a back-and-forth approach to reading text with chemical equations in Chapter 5, point out that they will want to take a similar approach when reading about mathematics.

Tell students that when they read text with math problems, they should plan to read the text more than once. The first read is a skim to get a general idea of what the text is saying. On the second read, they should focus on the math portion. As they read, they should look at the symbols and say to themselves what the symbols represent. For example, when they encounter the equation $F = m \times a$, they should remind themselves that the letter F means force, m is for mass, and a is for acceleration.

Sample problems require even more engagement. It is tempting to just skip over the lines of math, but this isn't effective for learning how to solve the problem. Students should start by comparing the top line of the problem with the second line. What has changed and why? If it's not obvious, they may need to check the text above or below the problem. Display this problem from "Of Kangaroos and Secretary Birds":

$$F = m \times a$$

$$F = 10 \text{ kg} \times 80 \tfrac{m}{s^2}$$

Kick Force

Say, "How did the author move from '$F = m \times a$' to '$F = 10 \text{ kg} \times 80\frac{m}{s^2}$'? To find out, glance back at the paragraph above the problem. It says, 'The scientist knows that the mass (m) is equal to 10 kg. The acceleration (a) is $80\frac{m}{s^2}$.' Aha, so the first step in solving this problem is substituting the given mass and acceleration into the equation. (Notice that there's an m here in the acceleration. This m stands for meters, unlike the m in the equation, which stands for mass."

Readers should keep comparing each line of the problem solution to the line before to ensure they understand what is being done. On the other hand, if they believe they have a general grasp of the problem, they can take a different approach. They can take a piece of paper and lay it under the first line of the problem and try to predict what the next line will be. Once they've made their prediction, they can move the paper to see if it is correct. If they are wrong, they should look around in the text to try to figure out how the author calculated the next line.

This can be a slow, and sometimes tedious, process, but learning how to learn from math texts is a critical skill that will serve them throughout their science and math education.

Journal Question

After students have completed the reading, give them the following prompt for their reading journals, which will help them internalize the strategy they practiced: A friend says she is stuck on her math homework. Make suggestions for how this friend could use the sample problems in her textbook to figure out how to do the homework.

Application/Post-Reading/Writing

- **Writing Prompt.** You experimented with launching pucks of different masses and different forces. Furthermore, you read about the formula relating mass, force, and acceleration ($F = m \times a$). Are the results from your experimentation and the sample graph provided consistent with the formula $F = m \times a$?
    - **Pre-Writing Suggestions.** Put the formula $F = m \times a$ on the board and have students review what each part of the formula stands for. Point out that in their experiments, the students did not measure acceleration exactly. However, when something accelerates more on a consistent surface, it will travel farther, just as the students' pucks did on the lab tables. Therefore, they can use distance traveled as a stand-in for acceleration. Ask them to consider what should happen, based on the formula, to

acceleration when force increases. Then ask if that is what they saw during their experiment. Likewise, ask what should happen, based on the formula, to acceleration when mass increases, and if that is what they saw in the experiment.
- o **Key Evaluation Point.** When force increases and mass stays the same, the formula predicts that acceleration (or distance traveled) should also increase. When mass increases and force stays the same, the formula predicts that acceleration (or distance traveled) should decrease. These results are consistent with the sample results given and should also be consistent with students' own exploration. Students may want to put sample numbers into the $F = m \times a$ formula to demonstrate the direction of the change in acceleration with changes of mass or force.
- **Thinking Mathematically.** Students will use the Here's the Kicker … worksheet (p. 118) to practice solving $F = m \times a$ problems based on experiments that have been done with animals.

Shuffleboard Hero

In games like shuffleboard, the objective is to get a weighted puck as near to a target on the other end of the board as possible. If you can do this over and over, you'll be hard to beat!

1. Spend a few minutes with your shuffleboard materials, trying to launch the puck (the plastic cap) a distance of exactly 40 cm. (Try launching it with and without some modeling clay in it.) What two variables could you change to affect how far the puck went?

2. Design an experiment to collect data on the effects of mass and force on how far the puck travels.

    a. First, think about force. How could you measure how much force you put on the puck?

    b. Describe how you could collect data on the movement of the puck when you change the force applied.

    c. How many times would you do each test described above? _____

    d. Now, think about mass. How could you measure how much mass the puck has?

    e. Describe how you could collect data on the movement of the puck when you change the mass of the puck.

    f. How many times would you do each test described above?

NATIONAL SCIENCE TEACHING ASSOCIATION

Figure S8.1. Shuffleboard Device

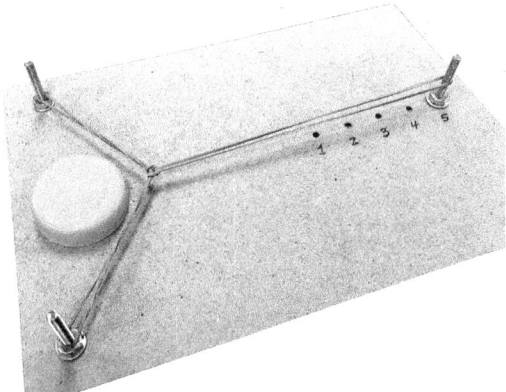

3. A group of students built a device to conduct an experiment similar to the one you have been working on. Their device is shown in Figure S8.1. To test the effects of mass, they experimented with one puck with a mass of 10 g and a second puck with a mass of 20 g. To look at force, they pulled the launcher back 1, 2, 3, 4, and 5 cm. They tested each combination three times and took the average of the distances moved. You can see their results in Figure S8.2.

Figure S8.2. Shuffleboard Test Results

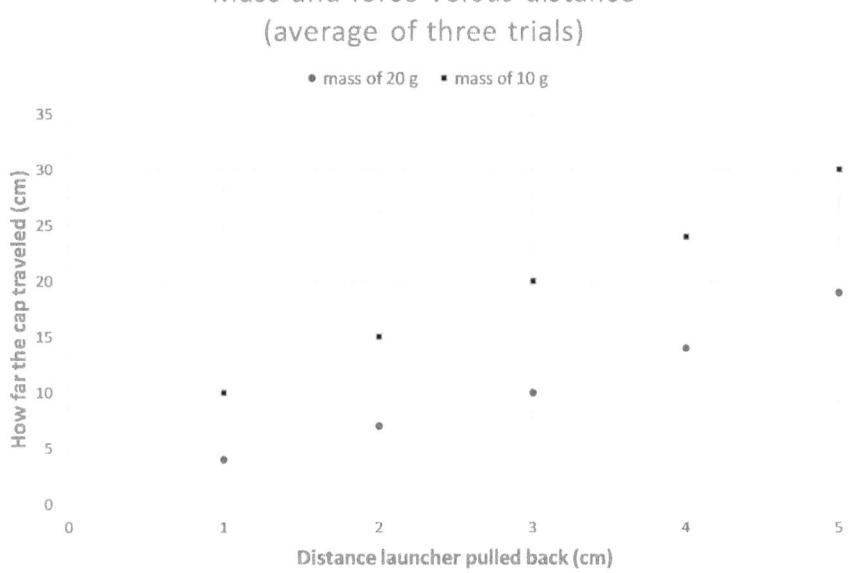

a. Based on this graph, if the mass increased but the force stayed the same, what happened to the distance that the puck traveled?

b. If the force increased but the mass stayed the same, what happened to the distance that the puck traveled?

Kick Force

4. You want to adjust your launcher for a competition. In that competition, you will be given a puck that weighs 10 g, and you want to launch it as close to 40 cm away as possible.

    a. Make a claim: How far back should you pull the rubber band to have your best chance at hitting the target distance? (Remember that you may have to make some changes to your launcher.)

    b. Explain how the data from the graph in Figure S8.2 supports your answer. (How did you calculate your response?)

Chapter 8

Of Kangaroos and Secretary Birds

Hopping kangaroos are a comical sight, but fighting kangaroos are even more bizarre. If two males are interested in the same female, they will swat at each other with their front paws, wrestle, and ultimately lean back onto their tails and give a fierce kick with both legs (Figure S8.3).

REMEMBER YOUR CODES
! This is important.
✓ I knew that.
X This is different from what I thought.
? I don't understand.

Figure S8.3. Kangaroo Battle

Secretary birds can also deliver a powerful kick (Figure S8.4). These four-and-a-half-foot-tall birds from sub-Saharan Africa eat snakes, including venomous snakes. Before they chow down, they protect themselves by stomping their snake snack to death.

Figure S8.4. Secretary Bird

Scientists who study animal movement often want to know how much force an animal's muscles can generate. Fortunately for them, Newton's second law of motion can help. Newton stated that force is equal to mass times acceleration. Or, in mathematical representation:

$$F = m \times a$$

Once Upon a Physical Science Book

Kick Force

Say a scientist wanted to measure the force of a kangaroo's kick. So the scientist trains a kangaroo to kick a 10 kg exercise ball. He finds that the ball accelerated at 80 $\frac{m}{s^2}$. How could the scientist use Newton's second law to calculate the force of the kick? The scientist knows that the mass (*m*) is equal to 10 kg. The acceleration (*a*) is 80 $\frac{m}{s^2}$. These values can be plugged into the equation as follows:

$$F = m \times a$$

$$F = 10 \text{ kg} \times 80 \frac{m}{s^2}$$

Solving the equation gives the following:

$$F = 800 \frac{kg \cdot m}{s^2}$$

The resulting unit of force is complicated. Force always includes mass, distance, and time squared. Those are a lot of units to specify! For this reason, scientists created a unit specifically for force and named it after Isaac Newton. One Newton (N) is equal to the force required to accelerate one kilogram of mass at 1 meter per second squared:

$$1N = 1 \frac{kg \cdot m}{s^2}$$

Therefore, the kangaroo can kick with a force of 800 N. Notice that this solution to the problem ignores friction. A ball rolling across the ground doesn't just have the force of the kick acting on it. It also has friction slowing it down. However, it is common to ignore the effects of friction when those effects are small and the calculation does not need to be exact.

Now let's look at a feathered kicker. Researchers studying secretary birds convinced a male bird named Madeleine to kick a rubber snake lying on a force plate, a device that measures forces. He smashed the snake with a force of about 200 N. Madeleine weighed about 4 kg. Assuming he put his whole body into the smash, what was his average acceleration?

Once again, scientists can use the formula $F = m \times a$ to get answers. Only in this case, they want to solve for acceleration, so they can rearrange the formula by dividing both sides by mass:

$$\frac{F}{m} = \frac{m \times a}{m}$$

$$\frac{F}{m} = \frac{\cancel{m} \times a}{\cancel{m}}$$

$$\frac{F}{m} = a$$

Then they can plug their data into the equation:

$$\frac{200 \text{ N}}{4 \text{ kg}} = a$$

$$\frac{200 \, \frac{\text{kg} \times \text{m}}{\text{s}^2}}{4 \, \text{kg}} = a$$

$$\frac{200 \, \frac{\text{m}}{\text{s}^2}}{4} = a$$

$$50 \, \frac{\text{m}}{\text{s}^2} = a$$

At the end of making this type of calculation, check the units. If the units are correct, then it is likely that the problem was set up correctly. In this case, $\frac{m}{s^2}$ are the correct units for acceleration. Based on this calculation, we know that a secretary bird's kick is $50 \frac{m}{s^2}$ — that's faster than the blink of an eye!

Newton's second law of motion is not only used to understand the motion of kicking birds. It's helpful for analyzing movement in many different situations, from sports to transportation to rocketry. In other words, researchers across the sciences rely on this simple formula to learn more about motion.

The Big Question

A young kangaroo kicked a ball with a force of 500 N. The ball had a mass of 2 kg. Write the equation and then put in the numbers to show how to solve for acceleration.

Thinking Mathematically: Here's the Kicker ...

Researchers have studied all kinds of kicks. Use the formula $F = m \times a$ to examine their results.

1. When horses kick, they can damage the objects around them. Researchers wanted to know how tough the objects around a horse needed to be. They found that horses could kick with a force of about 9,000 N. If the horse kicks a 90 kg bucket of water, how fast will the bucket accelerate?

   F = _____
   m = _____
   a = _____

2. Other researchers wondered how hard babies in the womb are kicking their mothers. They found that the little ones can kick with 45 N of force. If that baby could kick a 10 kg ball, how fast would it accelerate?

   F = _____
   m = _____
   a = _____

3. Researchers studied tae kwon do experts doing turning kicks. They had participants kick a roughly 50 kg sandbag outfitted with an accelerometer. The kicks had an average acceleration of about 13 $\frac{m}{s^2}$. How much force were the kicks producing?

   F = _____
   m = _____
   a = _____

4. Mantis shrimp do their punching with their mouthparts. They strike out and smash through the shell of the snail they plan to eat. Researchers determined that the mouthparts accelerate at about 100,000 $\frac{m}{s^2}$ and strike with 150 N of force. What is the mass of these tiny punching parts?

   F = _____
   m = _____
   a = _____

5. Would you rather be kicked by a tae kwon do fighter or a horse? Why?

6. Whose kick accelerates fastest—the tae kwon do fighter, the horse, or the mantis shrimp? Is it even close?

Chapter 9

Energy's Wild Ride

Topics
- Energy transformation
- Conservation of energy
- Potential and kinetic energy
- Energy efficiency

Reading Strategy
- Talk your way through it

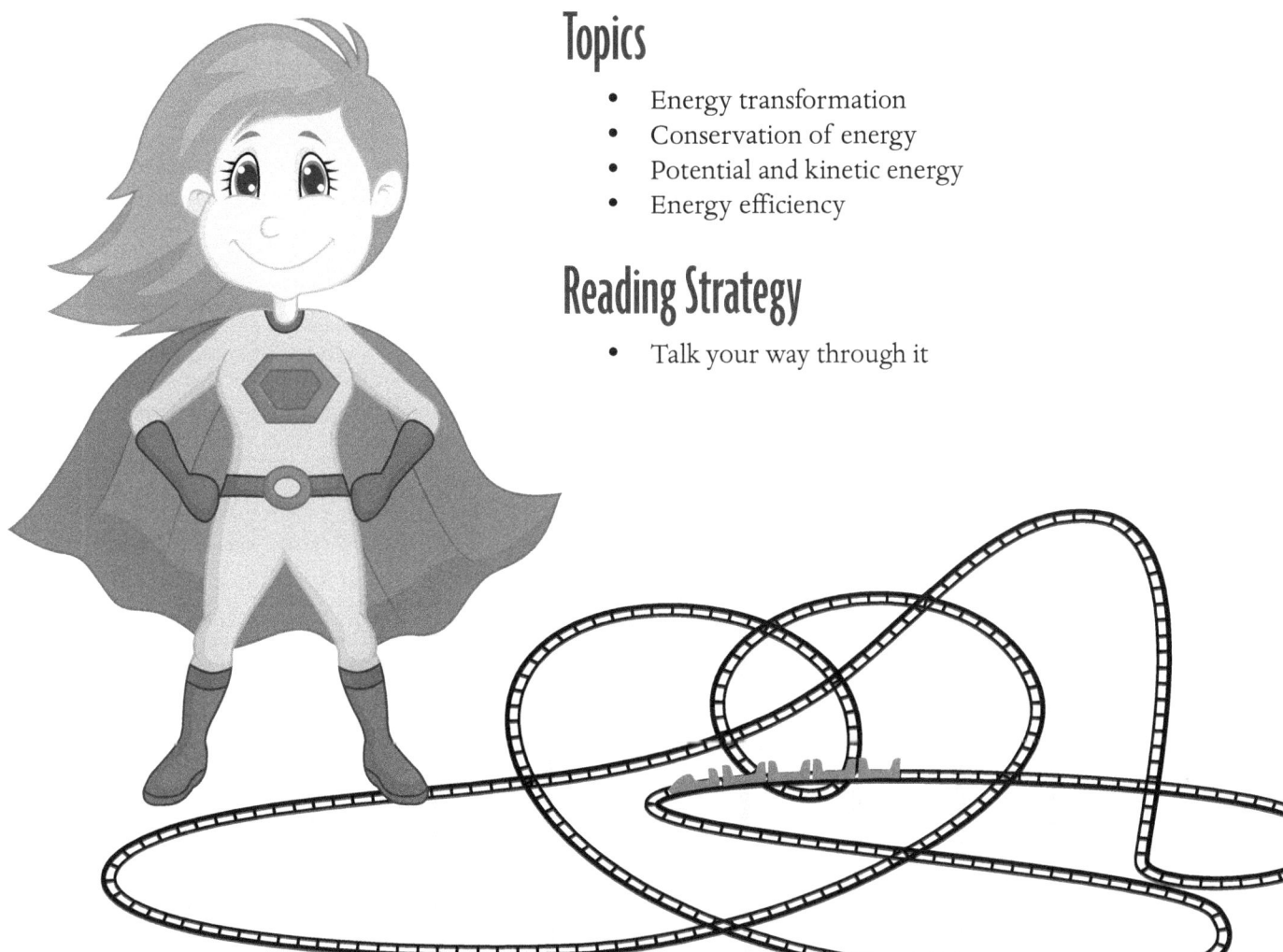

Connections to Standards

Next Generation Science Standards (NGSS) Correlations		
Standard **MS-PS3.** Energy (www.nextgenscience.org/dci-arrangement/ms-ps3-energy)		
Performance Expectation(s) The materials/lessons/activities outlined in this chapter are just one step toward reaching the performance expectations listed below. **MS-PS3-5.** Construct, use, and present arguments to support the claim that when the kinetic energy of an object changes, energy is transferred to or from the object.		
Dimension	**Element**	**Matching Student Task or Question From the Activity**
Science and Engineering Practice(s)	• Developing and Using Models • Engaging in Argument From Evidence	• Students use poppers and a rubber band–powered car as models to describe energy transfer through a system. • Students use data from the rubber band–powered car to explore energy transfer and information from the reading to form an argument that supports their claim that energy can be transferred to and from a system.
Disciplinary Core Idea(s)	PS3.A. Definitions of Energy • A system of objects may also contain stored (potential) energy, depending on their relative positions. PS3.B. Conservation of Energy and Energy Transfer • When the motion energy of an object changes, there is inevitably some other change in energy at the same time. PS3.C. Relationship Between Energy and Forces • When two objects interact, each one exerts a force on the other that can cause energy to be transferred to or from the object.	• Students use rubber band–powered cars to explore energy transfer and efficiency of the car model. • Students observe the energy transfer in a popper. • Students read about energy changes on a roller coaster, effects of friction, and energy efficiency. • Students describe the back-and-forth energy transfer in a roller coaster (from potential to kinetic and vice versa) in the Thinking Mathematically section.
Crosscutting Concept(s)	• Systems and System Models • Energy and Matter	• Students use poppers and a rubber band–powered car to examine energy transfer within a system. • Students read about the energy transfer through a roller coaster.
colspan Common Core State Standards (CCSS) Correlations		
Reading Standard(s)	• CCSS.ELA-Literacy.RST.6-8.2. Determine the central ideas or conclusions of a text; provide an accurate summary of the text distinct from prior knowledge or opinions.	• The reading skill for this chapter, talk your way through it, is a summarizing and remembering strategy.
Writing Standard(s)	• CCSS.ELA-Literacy.WHST.6-8.1. Write arguments focused on discipline-specific content.	• Students write an argument about which car is more energy efficient.

Chapter 9

Background

One of the crosscutting concepts in the *Next Generation Science Standards* (*NGSS*) focuses on understanding flows and cycling of energy and matter. In teaching energy transfer, we want to help students see the connections between energy in physical science and energy as it shows up in other disciplines. Energy is the same no matter where it exists—whether it's from the Sun, captured in photosynthesis, found in an earthquake or heated mass, stored in twisted rubber bands, or present in a moving roller coaster. In this chapter, students will explore energy change in toys and read about energy changes on a roller coaster.

Pre-Reading/Exploration

Materials for Activity

- Safety glasses with side shields or safety goggles
- 1–2 poppers (small half-balls that jump when inverted) per group
- 1 rubber band car per group (To build your own, see the materials list on p. 128.)
- Vinyl electrical tape
- Smooth floor or table

Activity

Use With Student Page(s): Energy on the Move (lab sheet) and Building Your Kinetic Kar (lab sheet)

For this activity, students will study two toys in which the mechanism of energy transfer is visible: poppers that jump after they are inverted and rubber band cars. Poppers can be purchased in bulk online or from party supply stores.

Part 1

Have groups turn their popper inside out and quickly place them back on the table in front of them. Using the poppers can be challenging for students at first, but they can usually get the hang of it within the first minute. Have students watch the popper's motion, and record their observations on their Energy on the Move lab sheets (pp. 125–127). Discuss with the class the answers to the questions from Part 1 of the lab sheet before moving on to Part 2. They may not figure out the path of the sunlight through plants to us, and they may not guess that energy from the popper

---

SAFETY NOTES
The following safety recommendations apply to all activities in this chapter:

- Wear safety glasses with side shields or safety goggles during the setup, hands-on, and takedown segments of the activities.
- Make sure the trajectory for the projectile is marked off and do not allow anyone to stand in its path.
- When using a glue gun, keep it away from combustibles or flammables in the work area. Never leave the glue gun plugged in if unsupervised.
- Do not touch the hot melted glue until it is fully dry and leave the glue gun unplugged for at least 20 minutes before touching the tip.
- Use caution in working with electrical sources to prevent shock, heat, or chemical burns.
- Any electrical power cords used near water must be plugged into a GFI-protected circuit. Presenters need to provide a GFI temporary power cord or power strip for this use.
- Any electrical cords should be off the floor. If that's not possible, they must be covered with tape or an electrical shield to prevent trip/fall hazards. Tape must be removed at the end of the session to prevent overheating wire fire hazards.
- Handle wires, skewers, and like objects with care as these sharp items can cut or scratch skin and eyes.

Continued

Energy's Wild Ride

SAFETY NOTES (continued)
- Use caution in dealing with heat-producing equipment, which can burn them!
- Appropriately dispose of lab materials at the end of the activities as directed by the teacher.
- Immediately report any lab accident to the teacher.
- Wash your hands with soap and water after completing the activities.

moves into air molecules, generating heat. Accept what they suggest at this time, but check back in after they read the article "Energy's Wild Ride" to ensure they make the connection. Collect all poppers before handing out the car supplies.

Part 2

The focus of the rubber band car activity is not on construction, although students will be more motivated if they have assembled the cars themselves. Instructions for building cars from common materials can be found on the Building Your Kinetic Kar lab sheet (p. 128). Some things to keep in mind: There will be a wide range of results based on the surface on which students run their cars. There will also be variation based on rubber band thickness and length. If shorter or thicker rubber bands are used, students may want to shorten the distance between their cars' axles. Ideally, a student's rubber band car shouldn't have tension on the band until it is being wound around the axle. If you prefer, you can shorten the lesson by building cars in advance, by having one class build cars and then reusing them with other classes, or by purchasing inexpensive rubber band car kits in bulk from a science supply company. (If you reuse the cars, simply remove the vinyl electrical tape for storage. The adhesive will migrate off the tape when left for long periods of time or at high temperatures.) You could also purchase premade rubber band cars—as long as the rubber band apparatus is fully visible for students to observe. Once students have their cars, they should practice using them and then answer the questions for Part 2.

Reading

Use With Student Page(s): "Energy's Wild Ride" (article)

Introduce the Reading. Tell students they are going to read more about rubber band cars and other things that require energy. Tell them to be looking to see if the guesses they made in their lab questions were correct.

Reading Strategy: Talk Your Way Through It

To introduce the strategy, display the following paragraph from the reading and have a student read it aloud:

> *Remember that energy is the ability of an object to do work. When an object such as a roller coaster train is moving, it's easy to picture how it could do work by, say, smashing into an empty cart blocking the track. The energy of an object in motion is called kinetic energy. But what about a train at the top of the first roller coaster hill, sitting still for just a moment before plunging down? What kind of energy could it have?*

That train has stored energy, called potential energy because it has the potential to do work. That train has the potential to race downhill.

Say, "There was a lot of information in that paragraph. Sometimes it's hard to remember everything when you read a lot of new information at one time. One way to remember more and make sure you understand what you're reading is to pause and talk yourself through what you just read."

Then model for students how to use the strategy of talk your way through it (also called pause, retell, and compare). For example, you might say, "Okay, so from the first reading of this, I remember that a roller coaster at the top of a hill has stored energy. I may not be exactly clear on how the energy is stored. I do remember there is a word for stored energy."

Ask the class to look back at the paragraph and see what else you should remember from it (and if anyone can explain what the article means by "stored energy"). Tell students that by pausing to talk yourself through dense passages, you can make sure you understand the current paragraph before moving on to the next one.

Other paragraphs in this article are also dense and may contain information students have not learned before, especially near the end of the article. Instruct them to try pausing after each section in the text to talk through (out loud, in their heads, or on paper) what they just read before moving on. If students are using reading groups, point out that a version of the talk your way through it strategy is built into the reading-group procedure when one member is asked to tell what he or she understands so far.

Journal Question

After students have completed the reading, give them the following prompt for their reading journals, which will help them internalize the strategy they practiced: Some people think that successful readers immediately "get it" whenever they read something. In fact, most successful readers use strategies like talk your way through it. A friend wants to know how you remember so much of what you read. How would you explain the talk your way through it strategy to your pal?

Application/Post-Reading/Writing

- **Writing Prompt.** Revisit your data from your rubber band car from before and after you taped the wheels. Make an argument to answer the following question: Which version of the car was more

Energy's Wild Ride

energy efficient? Use your data for evidence, and in your reasoning, trace the path of the energy as it moves through the car, using the terms *potential energy* and *kinetic energy*.

- **Pre-Writing Suggestions.** You may need to help students locate the data required for the writing assignment on their lab sheets (i.e., the charts they made and their data on what leads the wheels of their cars to spin out). Furthermore, you may wish to point them to the section of the article that talks about energy efficiency. Ask what signal words they could use to help them compare and contrast the two versions of the car (see Chapter 11 for suggestions).
- **Key Evaluation Point.** Students should argue that whichever car design traveled the farthest on the same number of twists is more energy efficient. For most students, this will be the car with vinyl electrical tape on the wheel; however, their data could differ. Students should explain that the energy was stored in the rubber band as potential energy and was released as kinetic energy as the car moved forward.

• **Thinking Mathematically.** Students use a diagram in the Up and Down the Gravity Road worksheet (p. 133) to examine changes in kinetic and potential energy.

Chapter 9

Energy on the Move

Part 1: Poppers

1. Take the popper toy your teacher has given you and turn it inside out (Figure S9.1). Quickly place it back on the table. What happened?

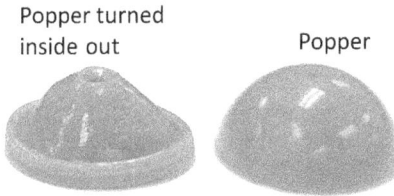

Figure S9.1. A Pair of Poppers

Popper turned inside out   Popper

2. It takes energy to jump. How did the popper get the energy to jump?

3. Think about your hands. It takes energy to move your fingers. Where did you get the energy you used to move your fingers?

4. Most energy on Earth originates at the Sun. How does energy from the Sun end up in your fingers? (Hint: Think about plants.)

Important Idea: Energy cannot be created or destroyed. It just changes form.

5. Take a guess: Where do you think the energy goes after the popper jumps?

6. The popper can jump even if no one is touching it. How does it store the energy until it jumps?

Important Idea: Energy can be stored. One way to store energy is in stretchy material.

Part 2: Kinetic Kars

For this section, use the premade Kinetic Kar provided for you or follow the instructions given by your teacher to construct a car.

1. Mark a starting line on a flat surface, allowing plenty of space to run your Kinetic Kar. While holding your car, gently turn the back axle two times to "wind up" the car's rubber band. Place your car on the flat surface at the starting line. Release the car. How far did it go?

2. Where did the car get the energy to move?

3. Take a guess: Where do you think the energy went after it powered the car?

4. Does changing the number of times you turn the back axle before releasing the car affect how far the car goes? Come up with a quick test to find out. Make a data chart in the space provided, and record your data:

Data Chart: Effect of Axle Turns on Distance

5. How does what you observed in Step 4 relate to energy movement? Complete the following statement and include the word *energy* in your response.

   As you increase the number of turns on the back axle, the distance the car travels *increases/decreases (circle one)* because _____

   _____.

6. When wheels lose their grip on the ground, the wheels turn, but the car does not move forward. This is called a spinout.
   - How many turns of the axle can you make before your car starts to spin out?
   - How far did the car go on that number of turns?

7. Cut a piece of vinyl electrical tape long enough to wrap around each back wheel. Wrap the wheels so they look like they have tires. Turn the back axle however many times caused a spinout in Step 6. Place your car on the flat surface at the starting line.
   - Did the wheels still spin out?
   - How far did the car go?

8. Repeat the same test you performed in Step 4. Collect your results, using the data to create a data chart in the space provided:

Data Chart: Effect of Axle Turns on Distance (Car With Vinyl Electrical Tape)

9. What is the advantage or disadvantage of adding the vinyl electrical tape to the back wheels?

Building Your Kinetic Kar

Materials (Per Car)

- 2 large wooden craft sticks

- 1 bamboo skewer

- 1 heavy bolt or used AA battery (for weight)

- 4 identical plastic drink caps (or two sets of two), with holes same size as the skewers pre-drilled in the center

- 1 straw

- 1 thin rubber band (1–2 mm in width and big enough to stretch 12 cm long)
  Note: If you use a shorter or thicker rubber band, you may want to shorten the distance between axles. Ideally, there should be no tension on the band until it is being wound around the axle.

- Vinyl electrical tape

- Scissors

- Ruler

- Low-temperature hot glue gun with glue sticks

- Safety goggles

> **SAFETY NOTE:** Even with a low-temperature glue gun, the glue and glue gun tip get hot enough to cause serious burns. Be careful not to touch the glue until it is fully dry and leave the glue gun unplugged for at least 20 minutes before touching the tip.

Assembly Instructions

1. Place one end of a craft stick on top of another to form a V. (See Figure S9.2.) Use the hot glue gun to join them at the place where they cross. The end where the sticks cross will be the front of the car.

2. Cut a 5 cm piece from one end of your straw. Glue the 5 cm piece to the connected part of your V-shaped craft sticks. Make sure the straw is positioned straight across this end of the V.

3. Take the remaining section of the straw and glue it across the opposite end of the V. Make sure that the two straw pieces are exactly parallel, like the lines on a sheet of notebook paper.

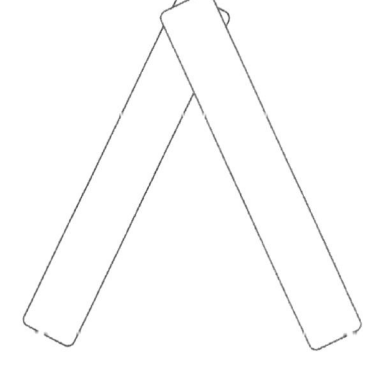

Figure S9.2. Step 1

These straws will keep your wheels in line with each other.

4. When the glue is dry (which takes about 10 seconds), trim out the middle of the long straw from your craft stick V. (Figure S9.3 illustrates Steps 2–4.)

5. Measure the bamboo skewer and use strong scissors or wire cutters to cut pieces of the skewer to the following lengths: 11 cm, 8 cm, 2 cm, and 3 cm.

6. Put the 8 cm section of bamboo through the straw piece at the front of the car and the 11 cm section of bamboo through the straw pieces at the other end of the V. These will form the wheel axles.

7. Attach the plastic caps (wheels) to the bamboo skewers with open end facing outward. Glue the wheels in place. **Before gluing, make sure the wheels are lined up to roll straight!** (Figure S9.4 illustrates Steps 5–7.)

8. Flip the car over so the wheels and axles are on the bottom. Glue the 3 cm skewer piece to the front of the car, pointing forward. Glue the 2 cm piece to the very center of the back axle.

9. Glue a heavy bolt (or other weight supplied by your teacher) to the craft stick next to a back tire.

10. Roll your car back and forth to make sure all parts are secure and balanced.

11. Loop the rubber band around the skewers on the front and back of your Kinetic Kar. (Figure S9.5 illustrates Steps 8, 9, and 11.)

Figure S9.3. Steps 2, 3, and 4

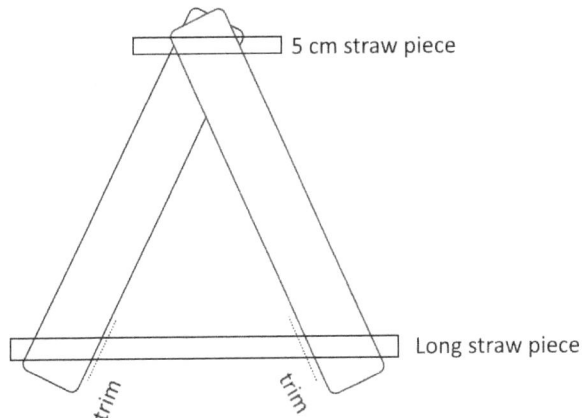

Figure S9.4. Steps 5, 6, and 7

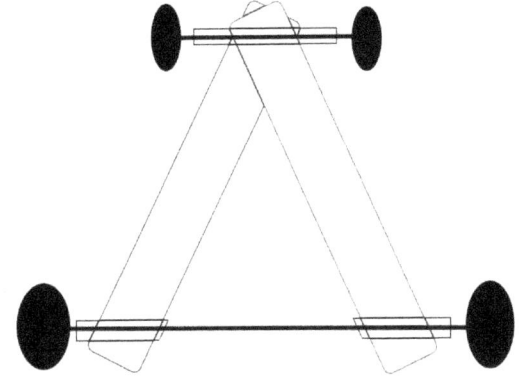

Figure S9.5. Steps 8, 9, and 11

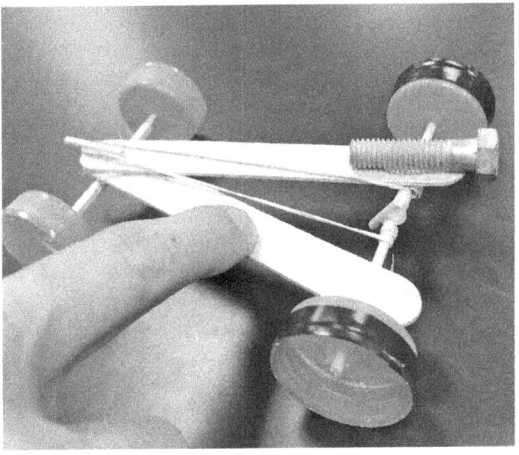

Energy's Wild Ride

Energy's Wild Ride

The year was 1885, and the first roller coaster with a cranking mechanism had just opened in Coney Island, New York. Riders of the Gravity Road gripped the sides of their wooden cars. *Clickety-clack. Clickety-clack.* The train inched up the long, long hill.

Its inventor, Phillip Hinkle, realized that the key to building an exciting ride was energy. You could put a lot of energy in to haul the train up, up, up one giant hill. (See Figure S9.6.) From there, gravity would carry the train through its swoops and sprints.

Remember that energy is the ability of an object to do work. When an object such as a roller coaster train is moving, it's easy to picture how it could do work by, say, smashing into an empty cart blocking the track. The energy of an object in motion is called *kinetic energy*. But what about a train at the top of the first roller coaster hill, sitting still for just a moment before plunging down? What kind of energy could it have? That train has stored energy, called potential energy because it has the *potential* to do work. That train has the potential to race downhill.

Roller coasters depend on a basic fact of physics: Energy cannot be created or destroyed. It only changes form. You can track those changes in a roller coaster as it moves. Kinetic energy in a motor hauls the train up a hill. By the time the train reaches the top, that energy has been stored in the train as potential energy. As the train heads downhill, the potential energy changes back into kinetic energy.

REMEMBER YOUR CODES
! This is important.
✓ I knew that.
X This is different from what I thought.
? I don't understand.

Figure S9.6. Climbing a Hill

Most roller coasters get their energy from one long pull up a hill.

Tracing Energy's Path

You can trace the energy moving through any kind of system. Take a rubber band–powered car, for example. You twist the rubber band with your hand. Your moving hand has kinetic energy. As the rubber band winds and stretches, it is storing potential energy in the rubber. You let go. The rubber band snaps back to its previous shape, changing that potential energy back into kinetic energy. Even after the rubber band stops moving, the car keeps going, carrying that energy even farther. As the car rolls, it rubs against air molecules and the table, passing some of the energy to them as friction.

Eventually, the car stops. Did the energy disappear? No way! You can't see it, but the car caused the air and table molecules to vibrate a little faster than they would normally. Scientists have a term for measuring how fast molecules are vibrating—it's called *heat*. If you run the car across the table enough times, you might even be able to feel the increase in heat with your hand. Heat is another form of kinetic energy because it comes from molecules in motion. Once energy has been passed on to the surroundings as heat, the energy still exists, but it is not generally useful for people anymore.

Sleek as a Submarine

Some objects channel their energy into useful work better than others. Picture a submarine sailing under the ocean. The submarine loses some of its energy to friction with water molecules (Figure S9.7). That's why submarines are rounded—to help the water flow by smoothly. The round shape limits how much energy is lost to the environment. In other words, the round shape makes the submarine more energy efficient. Energy efficiency is important in all kinds of devices, from vehicles to electronics to lighting. Energy-efficient objects save fuel, battery power, and electricity.

Figure S9.7. Friction Under Water

Submarines lose energy to friction with water molecules.

Energy in Living Things

Returning to the rubber band car, you could trace the energy movement even further back. Where did your hand get the energy it needed to twist the rubber band? It came from sugars stored in your muscle cells. Sugar is potential energy because energy is stored in the chemical bonds of the sugar molecule. Chemical energy can do work when the bonds are broken. And where did that sugar come from? Perhaps the strawberries you ate for breakfast, which came from a strawberry plant, which made the sugar using energy from the Sun.

Sleeker, Like a Shark

All animals get their energy from food. Take sharks, for example. They need food to power their muscles. Moreover, many types of sharks have to keep moving to

Energy's Wild Ride

keep oxygen flowing through their gills. If they stop swimming, they'll suffocate. As a result, they have evolved to be very energy efficient. Their bodies lose less energy to drag than any sort of vehicle people have been able to make. Remember the submarines mentioned earlier? Scientists would love to find a way to make these vehicles even more energy efficient. And engineers have been studying how sharks slip through the water in order to find ways to improve the energy efficiency of submarines.

As you go through your day, notice how energy is moving through the objects you use. You could be the person to make them more energy efficient! You might design a bicycle that goes farther with each push of the pedals. You might build a better rubber band car. Or perhaps, people will be screaming down the hill on *your* brand-new roller coaster.

The Big Question

Rub your hands together quickly 10 times. Feel the heat that forms. Starting with the Sun and ending with heat, trace the movement of that energy through your body. Label each type of energy as "potential energy" or "kinetic energy."

Chapter 9

Thinking Mathematically: Up and Down the Gravity Road

Pretend you are watching a roller coaster ride in which there is no friction. All the energy passes back and forth between potential and kinetic energy in the coaster car.

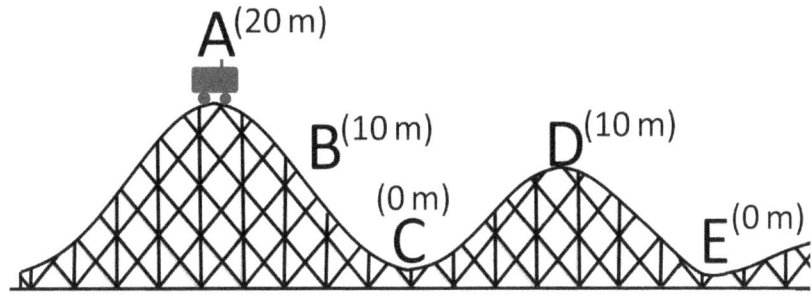

Assume the coaster car starts from rest at point A and rides across the track. Answer the following questions. *Some questions will have more than one answer!*

1. Where will the coaster car have the most potential energy?

2. Where will the coaster car have the least potential energy?

3. Where will the coaster car have the most kinetic energy?

4. Where will the coaster car have the least kinetic energy?

5. Where will the car have equal amounts of kinetic energy and potential energy?

6. Where will the car have the most total energy (kinetic + potential)? (Trick question—think carefully!)

7. Will the coaster car make it over the second hill? Why or why not?

8. If the second hill were 30 meters tall, would the coaster car get over the hill? Why or why not?

Once Upon a Physical Science Book

Chapter 10
Taking Your Temperature

Topics
- Heat
- Temperature
- Energy and particle movement

Reading Strategy
- Signal words for cause and effect

Connections to Standards

Next Generation Science Standards (NGSS) Correlations		
Standard **MS-PS1.** Matter and Its Interactions (*www.nextgenscience.org/dci-arrangement/ms-ps1-matter-and-its-interactions*)		
Performance Expectation(s) The materials/lessons/activities outlined in this chapter are just one step toward reaching the performance expectation(s) listed below. **MS-PS1-4.** Develop a model that predicts and describes changes in particle motion, temperature, and state of a pure substance when thermal energy is added or removed.		
Dimension	**Element**	**Matching Student Task or Question From the Activity**
Science and Engineering Practice(s)	• Developing and Using Models	• Students use a bean/marble model to describe the movement of "molecules" when different amounts of energy are added.
Disciplinary Core Idea(s)	**PS1.A.** Structure and Properties of Matter • Gases and liquids are made of molecules or inert atoms that are moving about relative to each other. • In a liquid, the molecules are constantly in contact with others; in a gas, they are widely spaced except when they happen to collide. In a solid, atoms are closely spaced and may vibrate in position but do not change relative locations. • The changes of state that occur with variations in temperature or pressure can be described and predicted using these models of matter. **PS3.A.** Definitions of Energy • The term "heat" as used in everyday language refers both to thermal energy (the motion of atoms or molecules within a substance) and the transfer of that thermal energy from one object to another. In science, heat is used only for this second meaning; it refers to the energy transferred due to the temperature difference between two objects.	• Students observe the different movement of food coloring molecules when placed in cold and hot water. • Students use a bean/marble model to view the molecular movement of water molecules when an additional substance (beans/marbles) is added. • Students observe a teacher demonstration that illustrates the spacing (density) and movement of molecules of hot and cold water. • Students measure an increase in volume when cold rubbing alcohol is heated, further illustrating the motion of atoms/molecules and the transfer of thermal energy from warmer to colder objects. • Students read about the history of the development of a thermometer and how it relates to atomic/molecular movement and heat.
Crosscutting Concept(s)	• Cause and Effect	• Students read about how temperature changes affect states of matter. • Students use evidence from a graph to show the relationship between temperature (cause) and the diffusion of coloring (effect) in the Thinking Mathematically section.
Common Core State Standards (CCSS) Correlations		
Reading Standard(s)	• CCSS.ELA-Literacy.RST.6-8.5. Analyze the structure an author uses to organize a text, including how the major sections contribute to the whole and to an understanding of the topic.	• The reading skill for this chapter asks for students to look for examples of cause-and-effect structures used in the text.

Continued

Chapter 10

| Writing Standard(s) | • **CCSS.ELA-Literacy.WHST.6-8.2.** Write informative/explanatory texts, including the narration of historical events, scientific procedures/experiments, or technical processes.
• **CCSS.ELA-Literacy.WHST.6-8.2.c.** Use appropriate and varied transitions to create cohesion and clarify the relationships among ideas and concepts. | • Students write an explanation of the nature of heat for a friend. In the explanation, they are specifically asked to use cause-and-effect transition words to make their explanations clear. |

Background

This chapter introduces the scientific concepts of heat and temperature. It will help students recognize the relationship between temperature and the movement of particles. As such, students should be familiar with the particulate nature of matter and the concept of density. While students encounter these two concepts regularly in science, many still struggle with them in middle school, and this lesson is an opportunity to reinforce those understandings.

Pre-Reading/Exploration

Materials for Activity

- Safety goggles
- Nitrile gloves
- Nonlatex aprons
- Warm and cold water (the greater the difference in temperature, the greater the effect)
- Red and blue food coloring with droppers
- 2 beakers per group
- 1 thermometer per group (Glass lab thermometers illustrate the idea better than digital probes.)
- 1 pan or plate with a rim per group
- 2 different colors of dried beans or marbles (enough to cover roughly half of each group's pan or plate)
- 2 identical wide-mouth glass bottles or drinking glasses with thick rims
- Index card or cardstock
- Chilled rubbing alcohol (Keep the bottle in an ice-water bath.)
- 5 or 10 ml glass graduated cylinders with corks or plastic wrap
  Note: If you only have plastic graduated cylinders, thin glass test tubes can be used instead. In this case, students will need to mark the liquid starting height with wax pencils or a piece of tape.
- 1 beaker with very warm water (hot water bath)

> **SAFETY NOTES**
> The following safety recommendations apply to all activities in this chapter:
> - Wear safety goggles, nitrile gloves, and nonlatex aprons during the setup, hands-on, and takedown segments of the activities.
> - Use caution when working with glass or plasticware, which can cut skin.
> - Never place food used in a lab activity in your mouth.
> - Review hazards on the Safety Data Sheets (SDSs).
> - Appropriately dispose of lab materials at the end of the activities as directed by the teacher.
> - Immediately wipe up any spilled liquid or pick up any materials on the floor.
> - Any electrical power cords used near water must be plugged into a GFI-protected circuit. Presenters need to provide a GFI temporary power cord or power strip for this use.
> - Use caution in handling hot water—it can burn skin.
> - Keep flammables (e.g., alcohol, food coloring) away from any active flames as they can cause fire or explode.
> - Immediately report any lab accident to the teacher.
> - Wash your hands with soap and water after completing the activities.

Once Upon a Physical Science Book

Taking Your Temperature

Activity

Use With Student Page(s): What Makes Heat Hot? (lab sheet)

Students will begin with several activities regarding the behavior of particles in hot and cold substances. Students can do Parts A and C, but Part B may work better as a demonstration.

Part A

In this activity, students will see that food coloring disperses more quickly in hot water than in cold water. They will then model that process using two colors of dried beans or marbles to represent molecules. Before class, set out materials for the students' models. Each group will need a pan or plate with dried beans or marbles to represent water molecules. There should be enough beans or marbles to cover roughly half the bottom of the pan or plate. Put a second color of beans or marbles in a separate bag or cup to represent particles of dye. You will want to have about ⅛ the number of dye particles as you have water molecules.

Part B

This is a common demonstration that allows students to observe that hot water floats on top of cold whereas cold water sinks into hot. You will fill one bottle with hot water and add a few drops of red food coloring. Then you will fill the other bottle with cold water and add a few drops of blue food coloring. Place the index card on top of the cold water bottle and flip the whole thing upside down. (Air pressure against the index card will prevent the water from pouring out, but you may wish to practice this step.) Set the cold water bottle on top of the hot one, lining up the mouths of the containers. Then pull out the card. The cold water will flow into the warm water, and the colors will mix. Repeat the process, but this time, place the cold water bottle on the bottom and the hot water bottle on top. The water will not mix. If you are uncertain of how to perform this demonstration, there are many videos and instructions online, including at the following site: *www.exploratorium.edu/snacks/inverted-bottles.*

Part C

In this step, students will measure the change in the volume of rubbing alcohol as it warms. To see this change of volume, the rubbing alcohol must be in a thin glass graduated cylinder and undergo a large temperature change. (*Note:* You must use glass; plastic containers expand and contract more, so they will not work.) In this case, students will place the graduated cylinders of cold rubbing alcohol into warm water baths. You can keep the rubbing alcohol cold by placing the bottle in a bowl of ice

Chapter 10

water. For the warm water, you can heat the water before class and keep it in a wide-mouth thermos, put warm water in a slow cooker and keep it set on warm, or place a beaker of warm water on a hot plate and keep the plate on the lowest setting. The hottest water available from the sink may not be sufficient to create a measurable change in volume.

Reading

Use With Student Page(s): "Feverish" (article)

Introduce the Reading. Tell students that they have been exploring how temperature changes affect matter. Now they are going to read about how those changes are used in a common household device.

Reading Strategy: Signal Words for Cause and Effect

Begin by displaying the following excerpt from the reading:

Mercury is a great liquid for thermometers because …

Ask students to predict what will be at the end of this sentence. They probably won't have any idea why mercury is a good liquid for thermometers, but you can point out that they can be pretty sure that the end of this sentence will give a reason. Ask which word from the excerpt signaled that "a reason is coming." Students should be able to see that the word *because* signals what the end of the sentence will hold. Put up the end of the sentence to confirm their predictions:

… it remains a liquid at most of the temperatures on Earth.

Tell students that certain words and phrases are signals for what the text is about to tell you. *Because* is a signal word for cause and effect. It's especially important in science reading to be on the lookout for words that signal cause and effect because so much of science involves figuring out the causes of things.

Show students the graphic organizer shown in Figure 10.1. Students can use such an organizer to help them think about and keep up with causes and effects in text. Have a student provide summaries of the cause and effect from the excerpted sentence about mercury. Point out that in this case, the cause comes later in the sentence than the effect!

Figure 10.1. Cause-and-Effect Graphic Organizer

Cause		Effect
	→	

Taking Your Temperature

Display this sentence next:

As a result, it was difficult for scientists to compare temperature information.

Ask if this sentence is giving the cause or the effect. (Answer: effect.) Then inquire about what word or phrase signaled that it was an effect. (Answer: *as a result*.) Ask if students think they should look for the cause in the sentence before or the sentence after this one in the text. (Answer: before.) Remark to students that the short phrase *as a result* gives all kinds of clues about the structure of the text they are reading. Finally, ask what other words might signal cause and effect. If students have difficulty responding, mention the words *because, cause/are caused by, consequently, as a result, therefore, thus, hence, for this reason, in response to*, and *leading to*.

You may also wish to tell students that cause and effect are not always indicated with a signal word. Display this sentence:

When energy is added to the particles, they move faster.

Help them identify cause and effect and observe that no signal words indicated the relationship in this case.

Have students watch for cause-and-effect signals as they read. It may be helpful to underline those text signals yourself before making printouts of the article and asking students to pause at each one to identify the cause and effect in the sentence.

Journal Question

After students have completed the reading, give them the following question for their reading journals, which will help them internalize the strategy they practiced: What is another school subject (besides science) where you learn about how some things cause other things? Do you think looking for text signals in that class would be helpful? Why or why not?

Application/Post-Reading/Writing

- **Writing Prompt.** A friend is drinking some hot chocolate and burns his tongue. He wonders why hot temperatures are painful. "It's not like heat can stab me or anything, " he says. "So how come hot things hurt?" Write an explanation to your friend, pointing out what temperature actually measures. Give at least one example, one cause or effect word, and one diagram.
    - **Pre-Writing Suggestions.** Ask what science vocabulary words students should include in their writing. Ask what writing words

they might use and help them see that the same cause-and-effect text signals they looked for while reading can also be useful in writing.
  - **Key Evaluation Point.** Students should say that temperature is the measure of how fast the particles in matter are moving. When hot liquid gets in the mouth, the particles transfer some of that energy to the tongue. Students may be able to extend that thinking to say that since the particles have energy, that energy can cause damage. This point, however, should not be required for a complete answer.
- **Thinking Mathematically.** Students use data from the Shaking It Up worksheet (p. 149) to determine how temperature affects the speed of molecular diffusion.

What Makes Heat Hot?

Part A: Temperature and Particle Motion

Experimentation

Start off by observing drops of food coloring in hot and cold water.

1. Fill one beaker with 250 ml of hot water. Fill the second beaker with 250 ml of cold water. Record the temperature of each beaker.

    Hot: _____   Cold: _____

2. Make some predictions: If you add three drops of food coloring to each beaker, what will happen? Will the food coloring spread out or stay in one place? Will anything different happen in the cold water versus the hot water? Why?

3. Add 3 drops of food coloring to each beaker at the same time. What happens to the color? Does the process move faster in one beaker than it does in the other?

Important Idea: Remember that water is made of lots of individual molecules that move around.

Modeling

You are going to use dried beans or marbles to model this molecular movement.

4. Your teacher will give you a pan with beans or marbles, which represent water molecules. You will also be given a second color of beans or marbles, which represent molecules of food coloring. Dump the second color of beans or marbles into one corner of the pan. Have one group member shake the pan back and forth rapidly. Have a second member time how long it takes until the "food coloring" molecules are roughly spread evenly among the "water" molecules. How long does it take for the "food coloring" molecules to spread evenly?

5. Pick out the "food coloring" molecules. Dump them back into a pile in one corner of the pan. This time, shake the pan slowly and time how long it takes

until the "food coloring" molecules are evenly spread out. How long does it take for the "food coloring" molecules to spread evenly?

6. Look back at your observations from the beakers of water. Based on this model, were the molecules moving faster in the hot water or in the cold water? Explain your answer.

7. This is only a model of molecular movement. It doesn't show exactly what happened in your beakers. In the pan, the molecules are only moving back and forth. How would that be different from water molecules moving around in a beaker?

Pick out the "food coloring" molecules and put them back in their container for the next group.

Part B: Temperature and Density in Water

Demonstration

Your teacher will perform a demonstration in which hot and cold water are combined. Watch the demonstration and answer the following questions:

1. Draw what happened when cold water was put on top of hot water. Label the colors as needed.

2. Draw what happened when the hot water was put on top of cold water. Label the colors as needed.

Taking Your Temperature

3. In a mixture, what sinks: the denser material or the less-dense material?

4. Based on this demonstration, are the molecules more densely packed in hot water or in cold water? How does the evidence from this demonstration support your claim?

Part C. Temperature and Density in Rubbing Alcohol

Experimentation

Observe rubbing alcohol in hot and cold water.

1. Put on safety goggles. Pour about 8 ml of cold rubbing alcohol into a 10 ml glass graduated cylinder. (If you have a 5 ml graduated cylinder, use 4 ml of alcohol.) Record your exact measurement here: _____
2. Cork the graduated cylinder or cover it with a small square of plastic wrap.
3. Fill a beaker about ¾ full of hot water. Place the graduated cylinder of alcohol into the hot water bath. Wait two minutes.
4. What is the volume of the alcohol in the graduated cylinder now that the alcohol is warmer? _____
5. The *number* of molecules of alcohol was the same when the alcohol was cool and when it was warm. What must have happened for the volume to increase?

Part D: Summarizing

What happens to the particles in a liquid when it is heated?

What does this movement do to the volume of the liquid?

Chapter 10

The drawing below shows a group of molecules on the left. Draw molecules in the empty space on the right to show what happens when heat is added.

Add Heat

Once Upon a Physical Science Book

Taking Your Temperature

Feverish

You wake up with a sore throat and a headache. "I'm sick!" you say. "I'm not going to school." A parent pops a thermometer under your tongue. Your temperature registers 102.5°F. You're definitely sick. You're headed to the doctor, not school.

People have known since ancient times that fever is associated with sickness. Doctors in the Roman Empire would feel a patient's skin and assign a category, such as "hot in the fourth degree." But the doctors were just guessing. One doctor's "hot in the fourth degree" might be another doctor's "hot in the second degree."

By the Middle Ages, both doctors and scientists had realized it would be useful to have a "ruler" they could use to measure temperature. To build it, they relied on a simple observation: The same amount of matter takes up more space when it is hot than when it is cold.

Thermo (Heat) Meters

Inventors made the first thermometers from thin tubes, which were marked with lines and set in larger tubes holding water or wine (see Figure S10.1). These devices often had a large, round top filled with air, which patients were sometimes instructed to hold in their mouths. The whole thing could be heated or cooled, causing the liquid to rise or fall accordingly.

At the time, thermometer makers didn't understand what was happening to cause the change in volume. Now we know that all matter is made up of particles, either atoms or molecules. These particles are always moving. When energy is added to the particles, they move faster. What we call "temperature" is a measure of the average speed of those particles. When you touch something that feels warm, you are feeling the movement of those particles colliding with the particles in your skin.

REMEMBER YOUR CODES
! This is important.
✓ I knew that.
X This is different from what I thought.
? I don't understand.

Figure S10.1. Early Thermometer

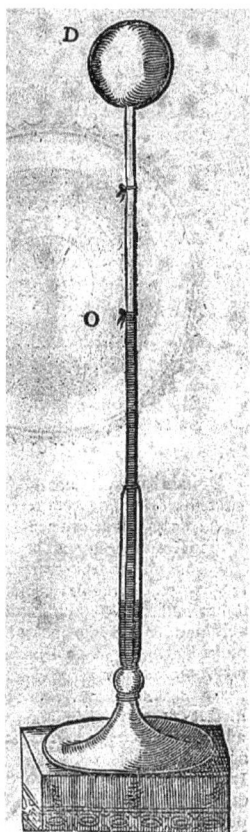

Heated particles move faster and bump into each other. As a result, they take up more space. Different liquids expand at different rates. In 1724, Dutch scientist Gabriel Fahrenheit selected mercury for his thermometer because it only takes a small change in temperature to create a noticeable difference in mercury's volume.

Going Through a Phase

The particles in a liquid flow and change positions, but they still hang together with nearby particles (see Figure S10.2). Suppose you kept adding energy in the form of heat to a liquid, such as mercury in a thermometer. Eventually the particles would move so fast that they would break loose from one another, forming a gas. Mercury reaches this *boiling point* at 356°C. Now suppose that instead of applying heat, you stuck the thermometer into a powerful freezer. The mercury would lose energy to the cold environment. The particles would slow down and link together, freezing and forming a solid. After freezing, the particles in the solid would jiggle in place without moving very much. Mercury reaches its freezing point at −38°C.

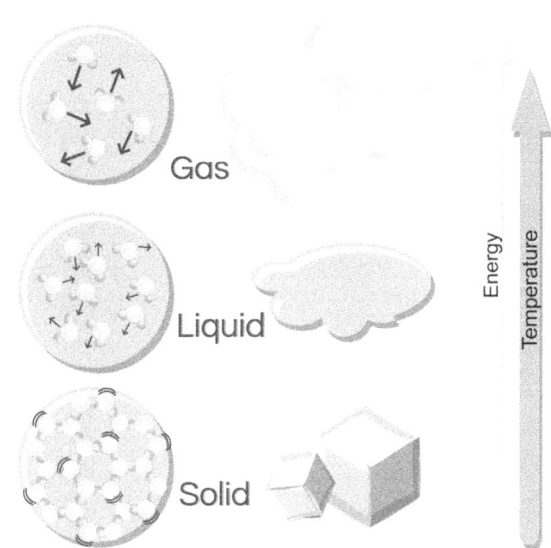

Figure S10.2. Models of Particle Movement in a Solid, Liquid, and Gas

A thermometer only works when the substance inside is in a liquid state. Mercury is a great substance for thermometers because it remains a liquid at most of the temperatures on Earth. Unfortunately, mercury is also extremely toxic, and glass thermometers have a tendency to break, which isn't a great combination for something you stick in your mouth. Therefore, modern glass thermometers usually substitute mercury for alcohol mixtures that are liquids at the temperatures in which the thermometer will be used.

Too Many Scales

In addition to his thermometer, Gabriel Fahrenheit devised a scale so that measurements would be consistent from one thermometer to the next. He used the freezing point of a saltwater mixture for the "zero" point on his scale. That meant that the freezing point of pure water ended up at 32 degrees and the boiling point ended up at 212 degrees. Those were awkward numbers to work with. So a few years later, Swedish scientist Anders Celsius came up with his own scale, setting the freezing point of water at 0 degrees and the boiling point at 100 degrees. Some scientists used Fahrenheit's scale. Others used the scale created by Celsius. As a result, it was difficult for scientists to compare temperature information.

Then the measurement situation got even more confusing. In the late 1700s, scientists learned that temperature was caused by moving particles. They realized that the lowest possible temperature would be at the point when particles stopped moving. They determined that particles would stop at −273°C and named this point

Taking Your Temperature

absolute zero. In response, yet another temperature scale was created. A scientist named Lord Kelvin shifted the Celsius scale down. He set "zero" on his Kelvin scale to equal −273°C, which meant that water would freeze at 273 Kelvin. You can see a comparison of all three temperature scales in Figure S10.3.

Figure S10.3. Comparing Temperature Scales

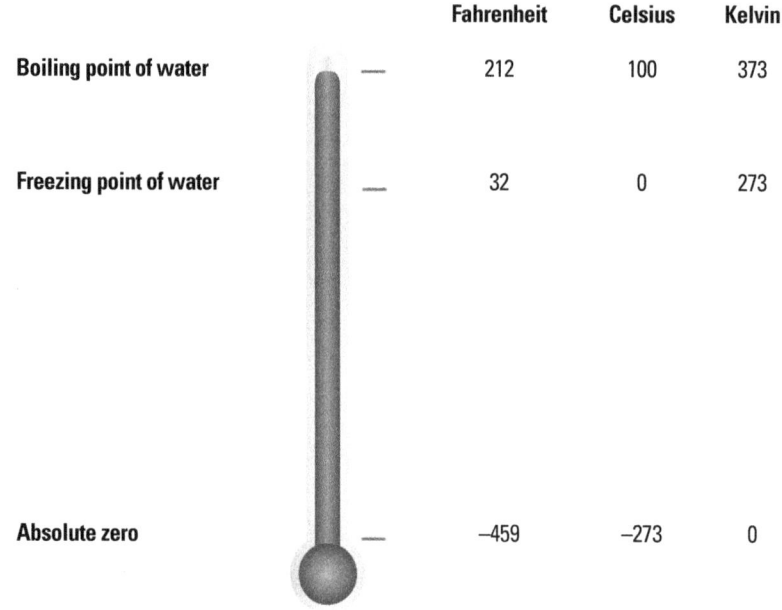

	Fahrenheit	Celsius	Kelvin
Boiling point of water	212	100	373
Freezing point of water	32	0	273
Absolute zero	−459	−273	0

To standardize the scales, modern scientists use Celsius, unless they are working with extreme temperatures, in which case they use Kelvin. The Fahrenheit scale is rarely used outside of the United States.

Next time you get sick and slide a thermometer into your mouth, think about how the particles in your body are trembling with energy. They transfer some of that heat into the thermometer. In response, the liquid increases in volume, and you can find out if you have a fever.

The Big Question

Think about the graduated cylinder filled with alcohol that you warmed during the lab. How is it similar to a thermometer? How is it different?

Thinking Mathematically: Shaking It Up

Hoyt, Hakim, and Hannah want to see how temperature affects the speed of molecules. They get three cups of water at different temperatures. Then they put three drops of food coloring in each cup. The graph in Figure S10.4 shows how long it takes the food coloring to completely diffuse, or spread evenly, through the cup.

Figure S10.4. Diffusion Graph

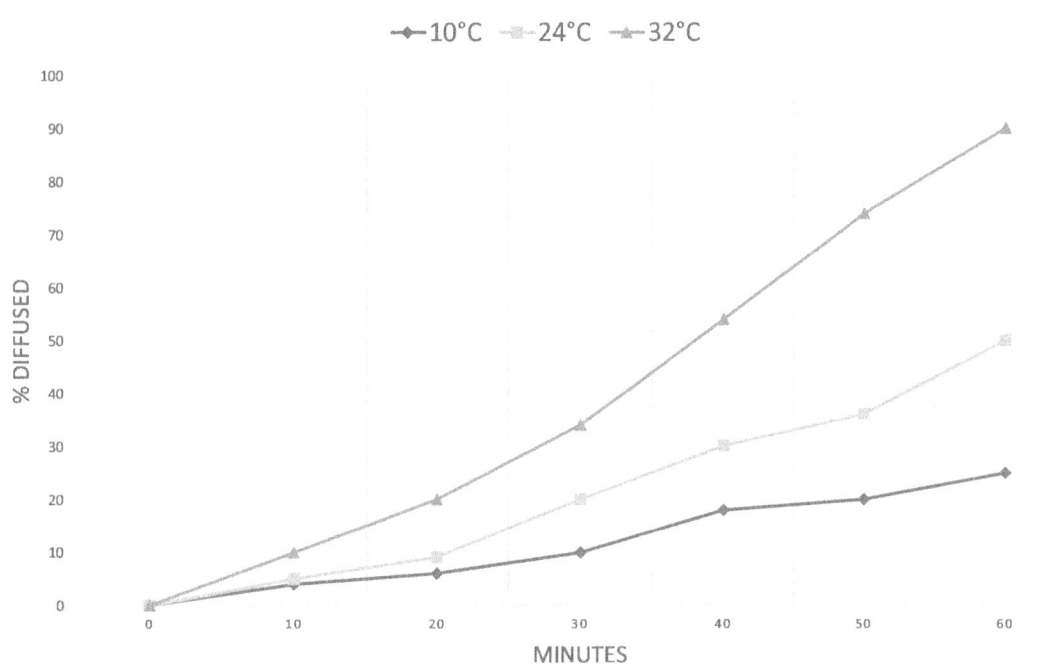

1. As time passes, what general trend do you see in the diffusion of the food coloring across all three cups? (circle one)

   The color molecules diffuse more.

   The color molecules diffuse less.

   The color molecules stay the same.

2. At 24°C, how long does it take for the food coloring to reach 50% diffusion?

3. What was the temperature of the water in the cup that was first to reach 20% diffusion?

Taking Your Temperature

4. What was the temperature in the cup that had the least diffusion after 60 minutes?

5. Make a claim about the effect of temperature on the time it takes for the coloring to diffuse in the water.

6. Use evidence from the graph to support your claim.

7. Use scientific reasoning to explain why temperature has an effect on the speed of diffusion.

Chapter 11
How to Not Die in Antarctica

Topics
- Heat transfer
- Insulation

Reading Strategy
- Signal words for compare and contrast

How to Not Die in Antarctica

Connections to Standards

Next Generation Science Standards (NGSS) Correlations		
Standard **MS-PS3.** Energy (*www.nextgenscience.org/dci-arrangement/ms-ps3-energy*)		
Performance Expectation(s) The materials/lessons/activities outlined in this chapter are just one step toward reaching the performance expectation(s) listed below. **MS-PS3-3.** Apply scientific principles to design, construct, and test a device that either minimizes or maximizes thermal energy transfer.		
Dimension	**Element**	**Matching Student Task or Question From the Activity**
Science and Engineering Practice(s)	• Constructing Explanations and Designing Solutions	• Students use data from their initial coat design investigation and evidence from the reading to design a coat for optimal warmth.
Disciplinary Core Idea(s)	**PS3.B.** Conservation of Energy and Energy Transfer • Energy is spontaneously transferred out of hotter regions or objects and into colder ones.	• Students investigate different types of coat insulators to slow the transfer of heat into an icy environment. • Students read about energy transfer and how it relates to the design of extreme cold weather gear. • Students calculate and analyze data on building materials to recommend the best building material to slow heat transfer in the Thinking Mathematically section.
Crosscutting Concept(s)	• Energy and Matter	• Students investigate different types of insulators to slow the transfer of heat into an icy environment. • In the Thinking Mathematically section, students calculate the R-value (a measure of how well a material blocks heat transfer) and cost of three different building materials to identify the best home insulator option for the price.
Common Core State Standards (CCSS) Correlations		
Reading Standard(s)	• **CCSS.ELA-Literacy.RST.6-8.5.** Analyze the structure an author uses to organize a text, including how the major sections contribute to the whole and to an understanding of the topic. • **CCSS.ELA-Literacy.RST.6-8.2.** Determine the central ideas or conclusions of a text; provide an accurate summary of the text distinct from prior knowledge or opinions.	• The reading skill asks students to look at how the author uses comparisons and contrasts to structure the article. • The big question asks students to summarize what the article says about the characteristics of good cold weather gear.
Writing Standard(s)	• **CCSS.ELA-Literacy.WHST.6-8.9.** Draw evidence from informational texts to support analysis, reflection, and research.	• Students use information from the text to reflect on their insulation experiments and decide what changes to make in their coats.

Chapter 11

Background

This chapter should be used near the end of the study of heat. Before doing this lesson, students should understand that temperature is a measure of the energy of particle motion and have had hands-on experiences with each of the three types of heat transfer: radiation, conduction, and convection. Here, students will look at stopping the transfer of heat through insulation. They'll "suit up" a warm water bottle to see if they can slow the transfer of energy into an icy environment. They'll read about coat design for extreme environments and then improve their coat design.

> **SAFETY NOTES**
> The following safety recommendations apply to all activities in this chapter:
> - Wear safety glasses with side shields or safety goggles during the setup, hands-on, and takedown segments of the activities.
> - Use caution when working with glass or plasticware, which can cut skin.
> - Appropriately dispose of lab materials at the end of the activities as directed by the teacher.
> - Use caution when working with ice cubes or hot water, which can burn skin.
> - Immediately wipe up any spilled water on the floor to avoid slipping or falling hazards.
> - Immediately report any lab accident to the teacher.
> - Wash your hands with soap and water after completing the activities.

Pre-Reading/Exploration

Materials for Activity (Per Group)

- Safety glasses with side shields or safety goggles
- 2 empty mini water bottles (8–10 oz.) with holes drilled in caps (¼ inch drill bit will create a hole that works for most classroom thermometers)
- 2 plastic sandwich bags for holding bottles and coat filling
- Choice of coat filling materials (whatever materials are available in your classroom setting, including aluminum foil, cotton batting or cotton balls, plastic beads, newspaper, shredded paper, packing peanuts, bubble wrap, corrugated cardboard, felt scraps)
- Tape for holding bag shoulders closed
- Warm tap water (roughly 40°C works well)
- Beaker or graduated cylinder of 250+ ml capacity
- 2 thermometers
- Stopwatch or clock
- Ice
- 4 quart-size plastic bags for holding ice
- 2 gallon-size plastic bags for keeping ice bags snug against the coat
- Freezer or cooler (optional)

Activity

Use With Student Page(s): Controlled Experiment: Bundle Up! (lab sheet)

Before class, prepare two "icy environments" for each group. For each icy environment, fill two quart-size bags with ice and place them inside a gallon bag. The "coats" will be wedged between the quart-size bags, as

How to Not Die in Antarctica

Figure 11.1. Coat in Icy Environment

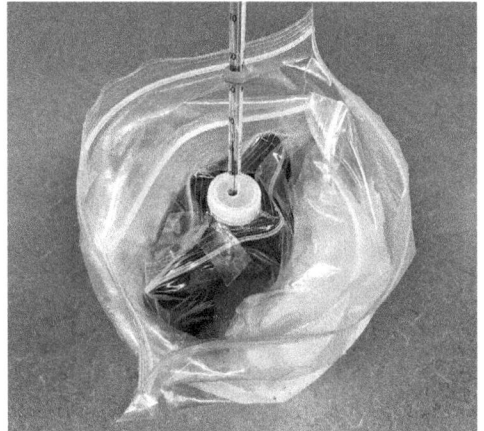

Figure 11.2. Coats Ready for Testing

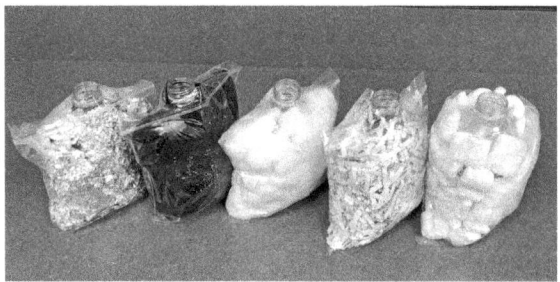

shown in Figure 11.1. The ice can be reused through several classes if you store the bags in a freezer or cooler in between.

Introduce the lesson by showing photos or videos of researchers working in Antarctica. The Weddell Seal Science multimedia page (*http://weddellsealscience.com/multimedia.html*) has a variety of options. Point out that it is hard for researchers to stay warm while working outside in extreme cold conditions. Hold up one of the small water bottles. Tell students that Captain Water Bottle is going to do some research in Antarctica and will need a good coat. The students must figure out how to make an effective coat that will keep the captain safe. To begin, they will need to gather data on the effectiveness of a variety of coat fillings.

Have each group of students select two coat filling materials they would like to test. To make it easier to reuse materials between classes, each group should use only one type of material per coat (Figure 11.2). If you find your classes are short on time, the coats can be premade and reused. If no one chooses a highly conductive material like foil, measure one yourself to add to the class data. Your class should have usable results in 20 minutes. Have students share their results with the entire class so each group has more data on more types of coat filling to use for their claim.

Reading

Use With Student Page(s): "How to Not Die in Antarctica" (article)

Introduce the Reading. Tell students that they are going to read about the insulation that researchers use in Antarctica to stay warm.

Reading Strategy: Signal Words for Compare and Contrast

To introduce the strategy, begin by displaying the following sentence, with blanks as shown:

Given that air transfers heat through convection, you might expect it to be a terrible insulator. However, _____ _____.

Ask students to predict what might go in the blank. Students usually guess something about air being good for insulation. Point out that they can make that guess even if they don't know what the word *insulator* means. Ask them which word gave them the biggest clue as to what should go in the blank. (Answer: *however.*) Put up the end of the sentence to confirm their prediction:

… it turns out that very small pockets of air are excellent insulators.

Signal words were first introduced with cause and effect in Chapter 10. If you haven't used Chapter 10 with your students, tell them that certain words and phrases are signals for what the text is about to tell you. The signal word *however* lets you know that two things are different, or that they contrast. When students see a signal word for contrasts, they should pause and ask themselves what two things are being contrasted and how they are different.

Ask students to list other words that might signal a contrast, and add these to the list as appropriate. Students may provide the following terms: *in contrast, on the other hand, conversely, whereas, alternatively,* and *instead*. The words *but, yet,* and *while* sometimes indicate a contrast.

Now add this sentence beneath the previous sentence, with blanks as shown:

In thick wool socks, the air is trapped between the cotton fibers. Similarly, a down coat _____.

Ask students to predict what might go in the blank. They may not know how a down coat traps air. But with coaxing, they can probably predict that the rest of the sentence will have something to do with trapping air. *Similarly* is a signal word for comparisons, or explanations of how two things are the same. When students see a signal word for comparisons, they should ask what things are being compared and how they are alike. Have students list other possible signal words and phrases for comparisons, including *similarly, in the same way, just like, just as, likewise, too,* and *also*.

Journal Question

After students have completed the reading, give them the following prompt for their reading journals, which will help them internalize the strategy they practiced. Write a short paragraph comparing and/or contrasting this year's science class with your class from last year. Use at least two signal words and underline them.

How to Not Die in Antarctica

Application/Post-Reading/Writing

Note: We strongly recommend asking students to revisit their coat filling materials and attempt to engineer a coat that will work better than those tested in the original experiment. However, engineering projects are beyond the scope of this book. If you choose this option, you may wish to edit the writing assignment to reflect their engineering experience.

- **Writing Prompt.** Pick the coat from the test with the best results and describe what changes you would make to it if you were using that material to make a coat for people. Why would you make those changes? Use information from the article "How to Not Die in Antarctica" (p. 159) to help you form your response.
  - **Pre-Writing Suggestions.** Ask students what information from the article helps explain the changes they would make. Have them think about what science words they want to include. Ask how they might integrate a sketch into their writing to help illustrate their ideas. (Students can use phrases such as *this picture shows* and *as you can see in the diagram.*)
  - **Key Evaluation Point.** Answers will vary, but students are likely to discuss at least one of the following topics based on the reading: air trapped in small spaces for insulation, water resistance and wicking, comfort of the coat, smell absorbency, or weight of the coat.
- **Thinking Mathematically.** Students will calculate the cost of insulation materials using the worksheet A Different Kind of Wolf (p. 162).

Chapter 11

Controlled Experiment: Bundle Up!

You are a textile engineer trying to select a filling material for a new coat. You need to find the best choice for helping Captain Water Bottle stay toasty for 20 minutes in an icy environment. You will be testing two coats, each with a different filling material.

1. To begin, look at the filling materials available. Decide with your group which two you want to test.

    a. Which two have you selected?

    b. Why did you choose them?

2. Prep your coats. The basic design has already been selected by another team at your company.

    a. Place a mini water bottle inside each of the plastic sandwich bags provided for you.
    b. Surround the water bottles with your choice of coat filling. Use only **one** type of filling in each coat. What variables will you need to control to make sure that the only difference between the coats is the filling material? Sketch your two coats here and label what you will do to keep them the same.

    c. Tape the sides of each bag closed to ensure the contents of your coat filling stay in the coat. The bottle opening should still be visible and above the edges of the seal.
    d. Carefully pour 150 ml of very warm tap water into each bottle.
    e. Screw the caps onto the water bottles. Insert the thermometers into the holes in the caps.
    f. Quickly measure the water temperature in each bottle and record it in the Water Temperature Data Table under Initial Water Temperature.

How to Not Die in Antarctica

3. Now it's time to test the effectiveness of each coat under cold conditions.

   a. Take your coats to the icy environment your teacher has prepared.

   b. Set a timer for 20 minutes. After 20 minutes, record the water temperature of each bottle.

Water Temperature Data Table

Coat Filling Material	Initial Water Temperature (°C)	Water Temperature After 20 minutes (°C)

4. Share the results with your class. Then use the class data to make a claim. Give the evidence and reasoning for your claim, as well.

   a. Claim: Based on our results, we recommend using _____ as the filling for the coat.

   b. Evidence (What data supports this claim?)

   c. Reasoning (Explain how your controlled experiment gives you confidence in your data. If you do not feel confident about your data, explain why.)

Chapter 11

How to Not Die in Antarctica

Weddell seals thrive in the ice and snow of Antarctica. Jay Rotella, a professor at Montana State University, heads the Weddell Seal Project that studies them. For two months each year, Rotella's research crew packs up and moves onto the ice to count, weigh, measure, and observe the seals (Figure S11.1).

REMEMBER YOUR CODES
! This is important.
✓ I knew that.
X This is different from what I thought.
? I don't understand.

Packing for this kind of camping trip is a big deal. Instead of tents, the researchers bring four small cabins made from metal shipping containers. They set the containers onto giant skis and drag them behind snowmobiles to a spot near one of the seal colonies they will study. For clothes, each researcher brings two duffle bags of ECW gear.

Figure S11.1. A Seal Pup Sighted by Rotella's Team

E-C-What?

ECW (extreme cold weather) gear is special clothing issued to every U.S. researcher who travels to Antarctica. The clothes are designed to help the scientists survive in this harsh environment. What materials work well for ECW gear? As you've probably observed, materials vary in how well they stop heat from moving. Lay your hand on a metal cabinet, and it feels cold. Touch a book in the same room, and it doesn't feel as cold. The cabinet and book are the same temperature—room temperature. However, heat moves through the metal cabinet more easily than it moves through the book. If you want to survive in Antarctica, you'd better pack materials that limit the flow of heat, called insulators.

Source: Jay Rotella, Weddell Seal Project, photo of Weddell seal was obtained under NMFS Permit 21158.

Getting Colder

The seal researchers lose some of their precious heat as radiation, or electromagnetic waves, carry heat away from the body. They can also lose heat by conduction. If a researcher grabs a metal drill with bare hands, for example, she will feel the heat moving into the drill. Although they lose small amounts of heat to radiation and conduction, the researchers lose the most heat through convection.

Figure S11.2. A Researcher Without a Hat Loses Heat Through Convection

To understand why convection is such a dangerous source of heat loss, imagine that a researcher heads into the Antarctic cold without a hat. His head is warm and heats the molecules of air nearby. Those molecules move faster, spread out, and float away, as shown in Figure S11.2.

Once Upon a Physical Science Book

Colder molecules rush in to fill the gap, and the process repeats. There is an endless supply of cold air to draw heat away, until the researcher puts on a hat—a puffy hat, with plenty of room for air.

Getting Warmer

Wait, a puffy hat with room for *air*? Given that air transfers heat through convection, you might expect it to be a terrible insulator. However, it turns out that very small pockets of air are excellent insulators. Air is not a good conductor. It also does not radiate heat very well. When air is trapped in small spaces, it cannot form efficient convection currents either. Insulating clothes are usually made from a material that radiates slowly and does not conduct heat well. That material is used to trap air.

For example, in thick wool socks, the air is trapped between the wool fibers. Similarly, a down coat traps air in goose or duck feathers (Figure S11.3) held between fabric layers. In foam, the pockets of air give the material a spongy feel.

When making ECW gear, another important factor to consider is water. Researchers' clothing can get wet from water in the environment. It can also get wet from perspiration. For instance, Weddell seal researchers can break into a sweat while slip-sliding after seal pups! Water leads to heat loss through both conduction and convection. Therefore, Antarctic researchers need fabrics that prevent water from the environment from soaking in and that draw sweat away from skin.

Figure S11.3. Feathers Trap Air Between Bits of "Fluff"

Cozy and Comfortable

The best ECW gear meets a few other requirements, as well. It should be comfortable, of course—who wants to deal with itchy or scratchy clothing? Furthermore, it should be lightweight. Every extra pound has to be hauled on long, tough walks across the ice. And, ideally, the cloth would not absorb odors. Researchers often need to wear the same clothes for days on end. Everyone on the expedition is happier if their clothes don't stink!

To achieve these goals, ECW clothes consist of at least three layers. The inner layer conducts as little heat as possible while wicking moisture away from the skin. It should also be soft and comfortable. By contrast, the outer layer is tough so that that it doesn't rip or wear away, and it has a coating to keep out rain and wind. The middle layer is usually the thickest and contains many pockets of air for insulation. (See Figure S11.4.)

Figure S11.4. Wearing ECW Gear in the Field

Source: Jay Rotella, Weddell Seal Project.

Rotella and his team may wear even more layers because the temperature can change dramatically while they are working. The morning may start with a windy snowmobile ride during which temperatures can hit −20°C. But by afternoon, the researchers may be lifting heavy seal pups in 20°C temperatures. On such a day, the team might start out wearing waterproof boots, long johns, insulated pants with a bib, a turtleneck, a fleece top, a thin down jacket, a work coat, a scarf-like neck warmer, sunglasses, a hat, and a hood. By afternoon, they may need to take off the jacket, the coat and hood, and even the neck warmer to avoid building up sweat that might turn icy later. Rotella and his team take the cold weather seriously and know the right clothes will keep everyone as warm and cozy as the seals they study.

The Big Question

Based on this article, describe the characteristics of good cold weather gear. Use the word *insulator* as part of your answer.

Thinking Mathematically: A Different Kind of Wolf

The story of the *Three Little Pigs* was about building strong houses to keep the wolf out. But real-life builders have another "big bad wolf" to consider: heat. Buildings need to keep heat out on a summer day and heat in on a winter day. This has more to do with putting in lots of insulation than building with strong materials.

Builders rate insulation with an *R-value*: a measure of how well a material blocks heat transfer. A higher number means the material is a better insulator. Take a look at the Building Materials R-Value chart and answer the questions that follow:

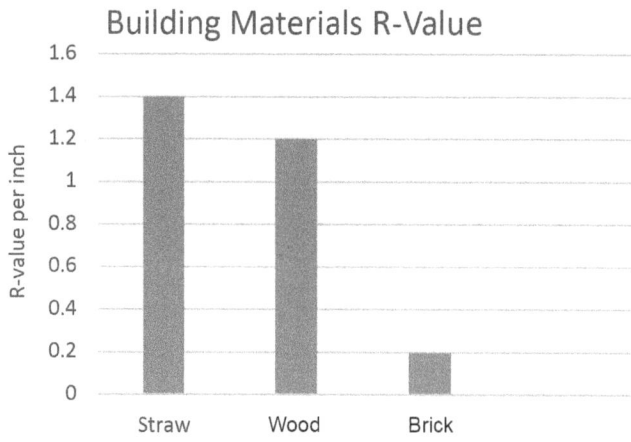

1. Which little pig home building material slows heat transfer the best?
2. How many inches of brick would it take to equal the heat-stopping power of one inch of straw?

Now you will choose what material—straw, wood, or brick—gives the most insulation for the best value. Selecting building materials involves a combination of different decisions, including considerations about cost. To make calculations easier, all materials will be sold in similarly sized units. A bale of straw (Figure S11.5) will be used to set the dimensions for each unit of material. This bale of straw is 18 inches thick. It will take about 16 boards or about 135 bricks to make a pile the same size as the bale of straw.

Figure S11.5. Bale of Straw

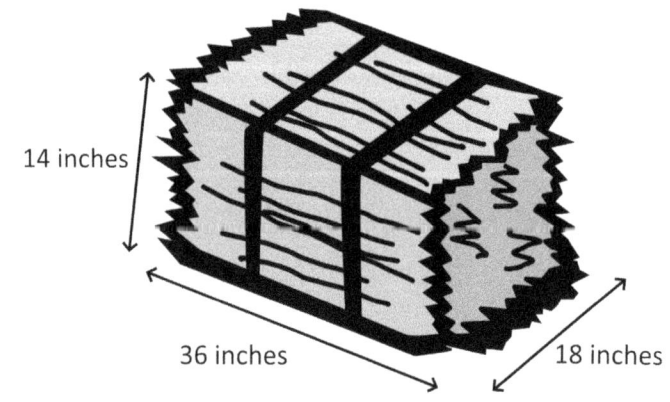

Chapter 11

Find the cost of each unit of material by multiplying the amount of material by the cost.

- 1 bale of straw × $5 per bale = _____ cost for a unit of straw

- 16 boards × $4 per board = _____ cost for a unit of wood

- 135 bricks × $0.50 per brick = _____ cost for a unit of brick

At 18 inches thick, what is the total R-value of each of the building materials?

- 18 inches of straw × _____ R-value per inch = _____ total R-value

- 18 inches of wood × _____ R-value per inch = _____ total R-value

- 18 inches of brick × _____ R-value per inch = _____ total R-value

- Which material represents the best value? _____

- What evidence would you use to justify this answer?

- A wall that is two straw bales deep would have an R-value of about 52. It would take three layers of wood to get the same result. How many units deep would a brick wall have to be to reach an R-value of 52?

- If you're trying to keep heat in or out of your house, which building material would you want to use?

Once Upon a Physical Science Book

Chapter 12
Ding-Dong Electromagnets

Topics
- Electromagnets
- Magnetic fields
- Doorbells

Reading Strategy
- Previewing diagrams and illustrations

Ding-Dong Electromagnets

Connections to Standards

	Next Generation Science Standards (NGSS) Correlations	
Standard		
MS-PS2. Motion and Stability: Forces and Interactions (www.nextgenscience.org/dci-arrangement/ms-ps2-motion-and-stability-forces-and-interactions)		
Performance Expectation(s)		
The materials/lessons/activities outlined in this chapter are just one step toward reaching the performance expectation(s) listed below.		
MS-PS2-3. Ask questions about data to determine the factors that affect the strength of electric and magnetic forces		
Dimension	**Element**	**Matching Student Task or Question From the Activity**
Science and Engineering Practice(s)	• Planning and Carrying Out Investigations	• Students test an electromagnet to demonstrate relationships between multiple independent variables and the dependent variable, magnet strength.
Disciplinary Core Idea(s)	PS2.B. Types of Interactions • Electric and magnetic (electromagnetic) forces can be attractive or repulsive, and their sizes depend on the magnitudes of the charges, currents, or magnetic strengths involved and on the distances between the interacting objects.	• Students design an electromagnet and test the magnitude of the force based on changes in currents and other possible variables.
Crosscutting Concept(s)	• Cause and Effect	• Students make an electromagnet to demonstrate the cause-and-effect relationships of variables that determine the strength of magnetic forces. • Students analyze data to predict the strength of an electromagnet given a higher voltage battery. • Students read about how the cause-and-effect relationships may be used to predict phenomena in natural or designed systems.
	Common Core State Standards (CCSS) Correlations	
Reading Standard(s)	• CCSS.ELA-Literacy.RST.6-8.7. Integrate quantitative or technical information expressed in words in a text with a version of that information expressed visually (e.g., in a flowchart, diagram, model, graph, or table).	• The reading skill for this chapter is helping students use diagrams and illustrations.
Writing Standard(s)	• CCSS.ELA-Literacy.WHST.6-8.1. Write arguments focused on discipline-specific content.	• Students make an argument about the effect of loops of wire on the strength of an electromagnet.

Chapter 12

Background

Electromagnets are used in many consumer and industrial applications. In this chapter, students will examine the features of electromagnetism through doorbells and sorting recycling. Prior to this chapter, students should have already studied magnetism and magnetic fields and be familiar with electric circuits.

Pre-Reading/Exploration

Materials for Activity

- Safety glasses with side shields or safety goggles
- 50–60 staples per group (separated)
- 6–10 toothpicks per group
- 6–10 scraps of torn paper per group
- 1 paper plate and 1 cup per group
- 1 permanent magnet per group
- 1 6-volt lantern battery per group
- 2 alligator clip test leads per group
- 1 long nail (about 50 mm in diameter) per group
- 28-gauge magnet wire (about 1 meter per group)
- Sheet of fine sandpaper, cut into roughly 5 cm squares
- Metric ruler
- Masking tape
- Scissors
- Compass (2–3 per class; can be shared among the groups)

> **SAFETY NOTES**
> The following safety recommendations apply to all activities in this chapter:
> - Wear safety glasses with side shields or safety goggles during the setup, hands-on, and takedown components of the activities.
> - Appropriately dispose of lab materials at the end of the activities as directed by the teacher.
> - Use caution when working with sharp objects (e.g., wires, nails), which can cut or puncture skin.
> - Use caution in working with batteries or other electrical sources to prevent shock and heat or chemical burns.
> - Do not keep batteries connected in a circuit for more than a few seconds—they can overheat and cause a fire or skin burns.
> - Immediately report any lab accident to the teacher.
> - Wash your hands with soap and water after completing the activities.

Activity

Use With Student Page(s): Recycling Quick Sort (lab sheet)

For this activity, students will first explore a situation in which electromagnets are useful: the extraction of iron-based metals from mixed recycling. Next, they will attempt to improve the basic electromagnet design by testing a variable to see if it strengthens the magnet.

Classroom electromagnets can be temperamental. The materials described here should give consistent results. However, you may wish to use different materials that you have on hand. Feel free to do so, but experiment ahead of time to ensure that your system works. This video shows instructions and tips for making the electromagnets: *www.youtube.com/watch?v=Wm9_DqQKmd0*.

> **SAFETY NOTE**
> The circuits students are building are essentially "short circuits"—the electricity flows straight back to the battery. Thus, they will get hot very quickly. Tell students to avoid leaving the circuits connected for more than a few seconds.

Ding-Dong Electromagnets

To start the activity, provide each group with a paper plate containing toothpicks, small paper scraps, and at least 50 staples mixed together, along with a paper cup labeled "collection bucket." Ask the groups to think of a way to quickly extract the staples for recycling. Someone is likely to suggest a magnet, particularly if you are in the middle of your unit on magnetism.

Give each group a permanent magnet and ask them to collect the staples and deposit them in the collection bucket. Tell them that they may not touch the metal themselves. They will realize that it is tricky to get the metal into the bucket, even with rubbing the magnet against the side of the cup. Point out that if they could turn the magnet on and off, they could avoid this problem. Provide groups with the batteries, wires, compasses, and nails and have students work through the Recycling Quick Sort lab sheet to create a simple electromagnet. Note that the compasses can be shared between groups as they are only needed for a short amount of time.

Reading

Use With Student Page(s): "When the Pizza Came" (article)

Introduce the Reading. Tell students they are going to read about a household object that is related to their recycling activity.

Reading Strategy: Previewing Diagrams and Illustrations

This strategy was first covered in Chapter 7. If you have not yet used Chapter 7 with your class, read pages 96–97 for information on how to introduce the strategy to students. Have students sit in pairs for this discussion. One student should be partner A and the other should be partner B. Lead the pairs through the following conversation, allowing one pair to share after each question:

Begin by displaying Figure S12.5 from "When the Pizza Came" (p. 174). Cover the title so that students focus on the visual elements.

- Have the A's describe to their partners what they see in this diagram. If they don't know the words for certain elements or what some parts of the diagram might represent, that's OK. Their main goal is to make observations.
- Have the B's make a prediction: What do they think the diagram has to do with the lab they just completed? Once again, they don't need to be correct. Affirm any reasonable response, but say they will soon have a chance to read an article that will give them more answers.

Now display Figure S12.7 (p. 174). Cover the title and caption.

- This time, have the B's describe what they see in the diagram.
- Have the A's predict what this diagram might have to do with electromagnets.

Finally, display Figure S12.3 (p. 173). Cover the title.

- Ask the B's to describe what they see to their partners.
- Have the A's make a prediction: How could this diagram relate to electromagnets?

Give students the article "When the Pizza Came" and tell them to check how accurate their predictions were as they read.

Journal Question

After students have completed the reading, give them the following question for their reading journals, which will help them internalize the strategy they practiced: Before reading, you looked at some diagrams, essentially taking these two steps: (1) You described what you saw in each diagram, without worrying about whether you knew the correct terms for different diagram parts, and (2) you made a prediction about what each diagram was trying to communicate. Which of these two steps was most helpful for you for understanding the text? Why?

Application/Post-Reading/Writing

- **Controlled Experiment.** Have the class look at Figures S12.6 and S12.7 in the article "When the Pizza Came." Then say, "Based on these images, make a prediction—do you think the number of coils in your recycling electromagnet matters? Would adding more coils change how much it could pick up?" Send students back to their Recycling Quick Sort groups. Have them use the Building Strength lab sheet (p. 176) to design an experiment to test their predictions. After you have approved their designs for safety, they should test them out.
- **Writing Prompt.** Use the questions at the end of your Building Strength lab sheet to make a claim and support it with evidence and reasoning.
  - **Pre-Writing Suggestions.** Ask students where they should look for evidence to support their claims. Ask what information from the article will help them explain their reasoning. Tell them to think about what science words they might want to include. Have students ponder what kind of diagram they would want to feature and what words they could use to tell the reader to

Ding-Dong Electromagnets

> **FIND OUT MORE**
> Show students how an electromagnet can pull a metal bar inside a doorbell at the Exploratorium website: *www.exploratorium.edu/snacks/magnetic-suction.*

look at the diagram. (Students can use phrases such as *this picture shows* and *as you can see in the diagram.*)

- ○ **Key Evaluation Point.** Ideally, students should be able to make the claim based on their results that more loops lead to a stronger electromagnet. They should give specific data from their lab to support their claim and explain that the magnetic fields created by the electricity moving through the loops overlap to create a stronger field. If a group's lab results support a different claim, lead them to dialogue with other groups to try to discover what led to the difference. In this case, the group should write up its own results, but note that other groups had different findings.

- **Thinking Mathematically.** Distribute the Using Data to Make a Prediction worksheet (p. 178). This activity demonstrates how data can be used for mathematical models and predictions. Point out to students that this data is "neater" than real-world data—it has been simplified to help them see the relationship between two specific variables. Once they see how this works, you can take any lab with directly proportional variables, have students eyeball a "best fit" line, and then see how closely their predictions match actual data.

Chapter 12

Recycling Quick Sort

You have a load of recycling, including paper, wood, and iron-based metals. You want to quickly pull out the metals and drop them into a collection bucket that can be sent to the furnace for recycling.

1. Brainstorm with your group. How could you quickly pull out the metals?

2. Try your idea. What new problem does this solution create?

Figure S12.1. Simple Electric Circuit

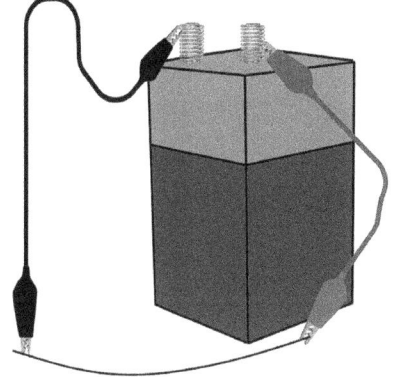

What if you had a magnet that you could turn ON and OFF? Let's make one. Use the materials from your teacher to assemble a basic electric circuit with a battery, alligator clips, and approximately 8 cm of wire, as shown in Figure S12.1. You will need to use sandpaper on both ends of your wire to remove the protective coating so that the alligator clips can transfer electricity. After you have the circuit, disconnect one wire to the battery so that it doesn't overheat.

> **SAFETY NOTE:** The circuits you are building today are essentially "short circuits." They will get hot very quickly. Do **NOT** leave the circuits connected for more than a few seconds.

Set the compass on top of the wire in your circuit. Remember that a compass points in the direction of a magnetic field.

3. In what direction is your compass pointing?

4. Connect your circuit. What happened to the compass when you connected the circuit?

Once Upon a Physical Science Book

Ding-Dong Electromagnets

5. Based on the compass's behavior, do you think the wire has a magnetic field when electricity is running through it? Why or why not?

6. Can your wire-magnet pick up staples? If so, how many can it lift?

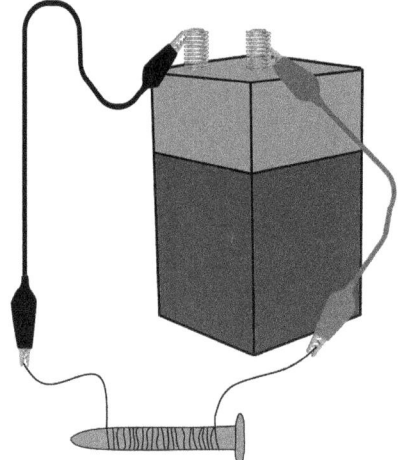

Figure S12.2. Circuit With Loops

The magnetic field on a straight wire is weak. You can make it stronger by looping the wires around a nail. Disconnect your wire. Take a longer piece of wire (about 40 cm long) and use sandpaper to remove the coating from the ends. Wrap the wire around the nail at least 20 times. Make sure to leave a little unwrapped wire on each end so you can connect it to the alligator clips, as shown in Figure S12.2. Use a small piece of masking tape to hold the wrapped wire in place, and then clip the wire to the battery.

7. Connect the wires to the battery. Can this device pick up staples? If so, how many can it lift?

This type of device is called an electromagnet because it uses electricity to create a magnet. Use your electromagnet to pick up some metal recycling and drop it in the collection bucket.

8. What do you need to do to your electromagnet to get the recycling to fall into the bucket?

Chapter 12

When the Pizza Came

"Delivery from Pickle Pumpkin Pizza Company!" calls out the pizza delivery person. He reaches for the doorbell and presses the button. *Ding-dong!*

Freeze Frame 1

Let's slow that down. The delivery person reached for the doorbell. When he pressed the button, it connected two wires and allowed electricity to flow through the system, as you can see in Figure S12.3. Electricity ran to the doorbell and set off the chime. *Ding-dong!*

> **REMEMBER YOUR CODES**
> ! This is important.
> ✓ I knew that.
> X This is different from what I thought.
> ? I don't understand.

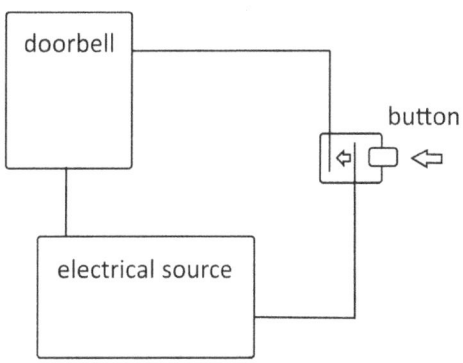

Figure S12.3. Basic Doorbell Circuit

Freeze Frame 2

Let's slow that down again and peek inside the doorbell. A traditional doorbell usually has two chimes that look like the slats on a xylophone. The shorter one makes the *ding* sound. The longer one makes the lower-pitched *dong* sound. There is a steel or iron striker between the chimes, as you can see in Figure S12.4. When electricity flows through the doorbell, the striker knocks against the short chime. Then a spring pushes it back to hit the long chime. *Ding-dong!*

Figure S12.4. Inside a Doorbell

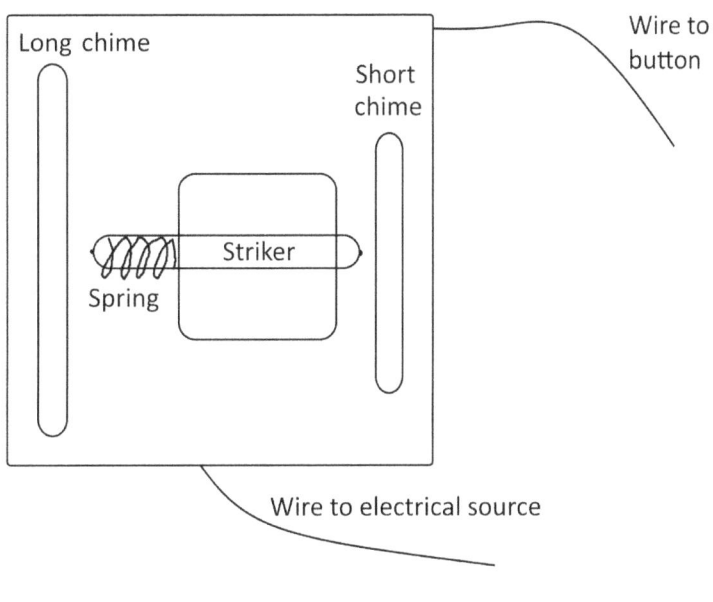

Once Upon a Physical Science Book

Ding-Dong Electromagnets

Freeze Frame 3

Let's slow that down even more. Why does running electricity through the doorbell move the metal striker? In the early days of studying electricity, scientists discovered a strange relationship. Moving magnetic fields could create electricity, and moving electricity could create magnetic fields. They found that if they ran electricity through a wire or other straight piece of metal, it would create a temporary magnetic field, as shown in Figure S12.5. The doorbell uses a magnet generated by electricity to jerk the striker backward against the chime. *Ding-dong!*

Freeze Frame 4

Hold on. It would seem like the doorbell maker would want to run electricity through the striker to magnetize it. However, a straight piece of metal or wire carrying electricity forms a very weak magnetic field. It's barely strong enough to move tiny iron filings, much less the striker on a doorbell. Stronger magnetic fields can be created by twisting the electrified wire into a loop. You can imagine it like a curled Slinky (Figure S12.6). On the inside of the loop, the coils of the magnetic field are very close together.

Adding additional loops provides more magnetic fields, which overlap to build strength. The fields merge to create one large magnetic field, as shown in Figure S12.7. When electricity passes through the looped wire in a doorbell, it attracts the striker and draws it back to hit the chime. *Ding-dong!*

Figure S12.5. Electricity in a Wire Produces a Magnetic Field

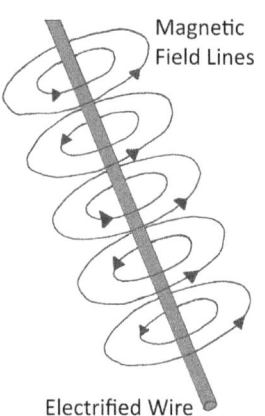

Figure S12.6. Magnetic Field of a Curved Wire

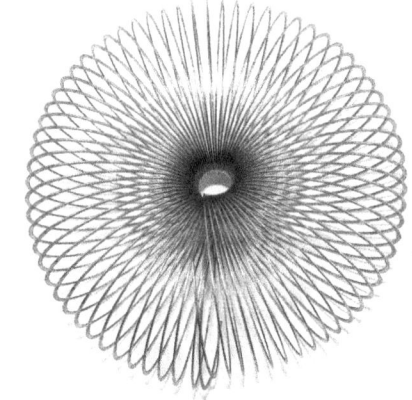

Inside a curved wire, the magnetic field comes together like a curled Slinky.

Figure S12.7. Magnetic Field in Loops

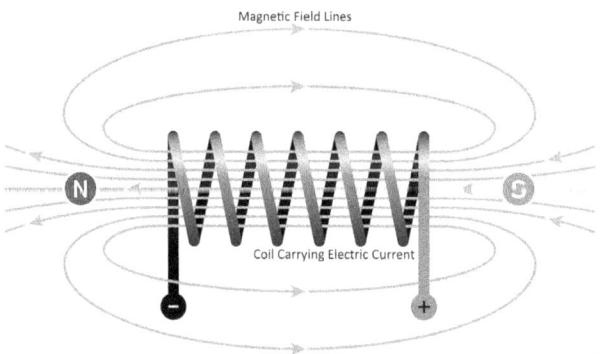

Loops cause magnetic fields to overlap and get stronger.

Chapter 12

Freeze Frame 5

Let's slow that down one more time. The striker doesn't just move into the magnetic field of the looped wire. It becomes a magnet while it's there. When a metal containing iron moves into a magnetic field, the electrons in the iron align themselves to match that magnetic field. Now the wire's magnetic field overlaps the striker's magnetic field, making the overall magnet even stronger. A magnet formed from loops of wire around an iron core is called an electromagnet. *Ding-dong!*

Fast-Forward

The delivery person presses the button, completing the circuit and sending electricity to the doorbell (Figure S12.8). The electricity flows through the coils of copper wire, forming a strong magnetic field, which attracts the striker and pulls it into the center of the coils. The striker magnetizes, which increases the overall strength of the electromagnet, as it hits the small chime. The delivery person releases the button, which stops the current, turns off the magnet, and releases the magnet's hold on the striker. The spring then pushes the striker back against the long chime. *Ding-dong!* Another perfect Pickle Pumpkin Pizza pie delivered.

Figure S12.8. Doorbell

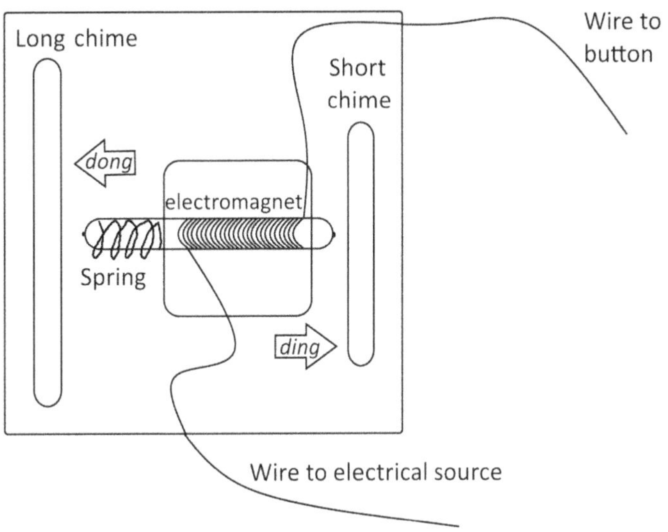

The Big Question

Describe why a looped wire makes a better electromagnet than a straight wire.

Once Upon a Physical Science Book

Building Strength: Improving Electromagnet Design

The recycling center needs an electromagnet that can quickly collect large amounts of iron and steel and move it to the collection bucket. Your group needs to figure out how to make your electromagnet stronger to meet this need.

> **SAFETY NOTE:** The circuits you are building today are essentially "short circuits." They will get hot very quickly. Do **NOT** leave the circuits connected for more than a few seconds.

What change in the design (variable) will you test? _____

Draw and label your experimental setup.
How many times will you repeat the test for each experimental design?
What variables will you need to keep the same to have a fair test?

NATIONAL SCIENCE TEACHING ASSOCIATION

Chapter 12

Specifically, what will you measure to compare the experimental and control groups? (You may want to use phrases like *how many, how far, how much,* or *how long* in your answer.)

SAFETY NOTE: Do **NOT** conduct your experiment until your teacher has approved your design.

Data: Make a data table to collect your results.

- **Make a claim:** Based on your results, how does the number of loops affect the strength of the electromagnet?

- **Evidence:** What data supports your claim?

- **Reasoning:** Use information from the article "When the Pizza Came," along with data from this experiment, to explain why and how the number of loops affects the strength of the magnet. Include a diagram in your answer.

Once Upon a Physical Science Book

Ding-Dong Electromagnets

Thinking Mathematically: Using Data to Make a Prediction

The voltage running through an electromagnet affects the strength of the magnet. A group of students did an experiment with an electromagnet with 50 loops. They connected it to batteries of various voltages and then lifted up paper clips to test for strength. They wanted to find out how strong the magnet would be with a 9-volt battery but none were available. Let's look at how they can use the data they *do* have to predict what might happen if they had a 9-volt battery.

Voltage Data Chart

Battery Volts	Average Number of Paper Clips Lifted
1.5	12
3.0	24
4.5	36
6.0	48

Make a line graph of the data:

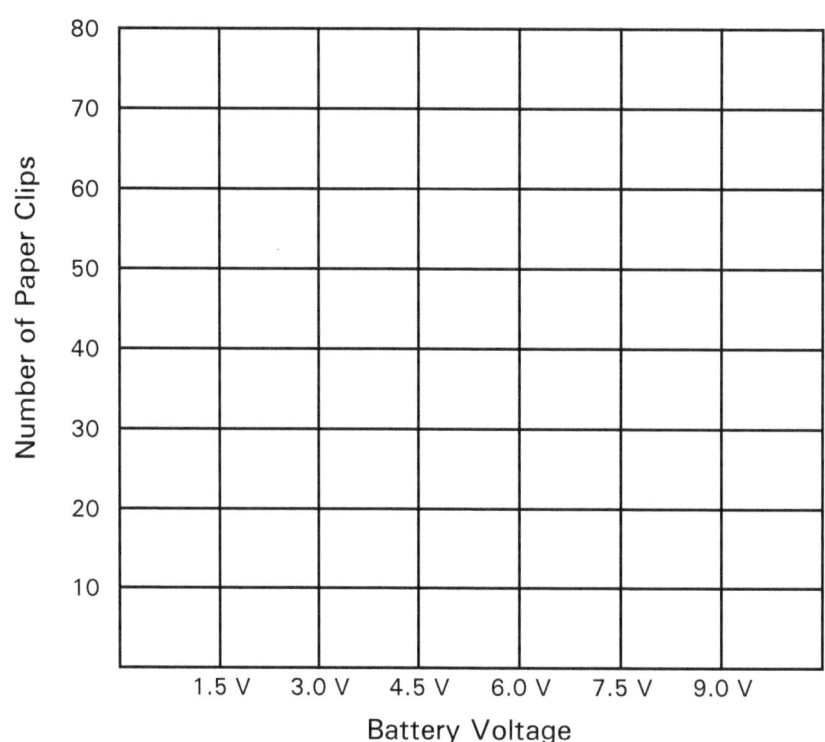

Chapter 12

Using Data to Make Predictions

1. Connect the dots you have plotted. Does the data make a straight line or something close to a straight line?

When the relationship between two variables makes a straight line when plotted on a graph, the variables are said to be "directly proportional." If the variables are directly proportional, you can predict other values by extending the line.

2. Take a ruler and place it against your line. Extend the line to the *x*-axis location for 9 volts. How many paper clips do you predict the students could lift with a 9-volt battery?

3. Without testing the 9-volt battery, can the students be certain that their prediction is correct? Why or why not?

Once Upon a Physical Science Book

Chapter 13
All About Bat Waves

Topics
- Introduction to waves
- Wave vocabulary
- Bat echolocation

Reading Strategy
- Finding the meaning of new words

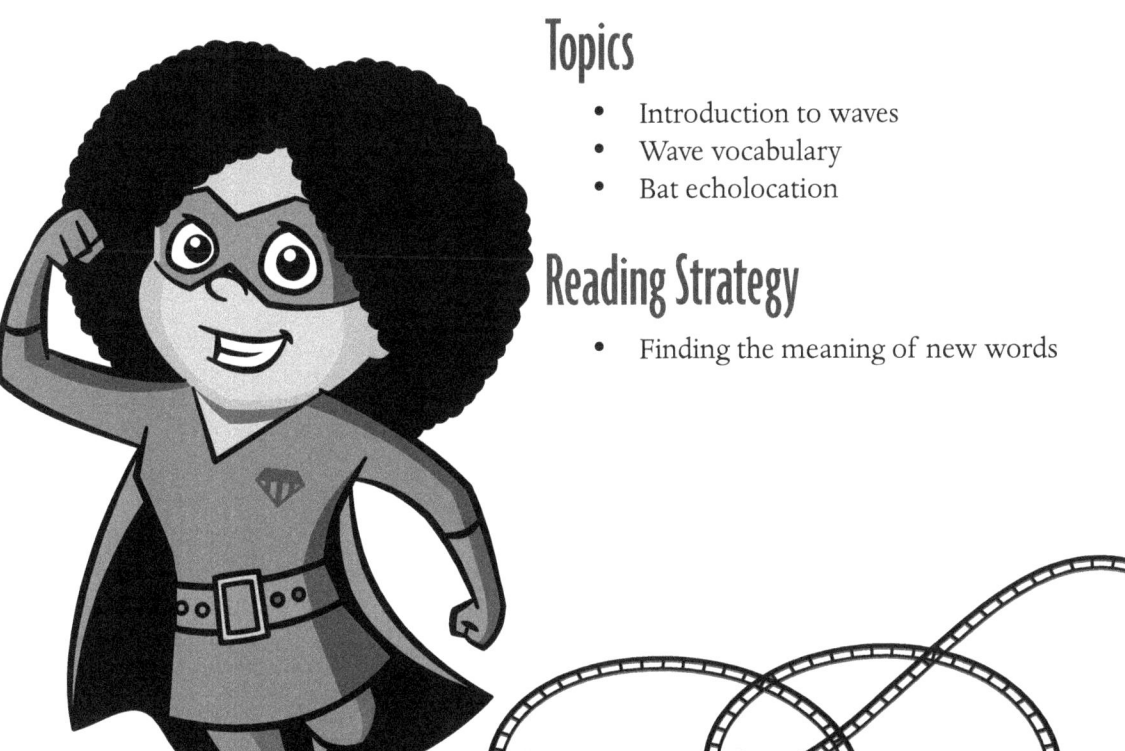

All About Bat Waves

Connections to Standards

Next Generation Science Standards (NGSS) Correlations		
Standard		
MS-PS4. Waves and Their Applications in Technologies for Information Transfer (*www.nextgenscience.org/dci-arrangement/ms-ps4-waves-and-their-applications-technologies-information-transfer*)		
Performance Expectation(s)		
The materials/lessons/activities outlined in this chapter are just one step toward reaching the performance expectation(s) listed below. MS-PS4-1. Use mathematical representations to describe a simple model for waves that includes how the amplitude of a wave is related to the energy in a wave. MS-PS4-2. Develop and use a model to describe that waves are reflected, absorbed, or transmitted through various materials.		
Dimension	**Element**	**Matching Student Task or Question From the Activity**
Science and Engineering Practice(s)	• Mathematics and Computational Thinking • Analyzing and Interpreting Data	• Students infer the relationship of frequency and amplitude to wave (energy) output in the Thinking Mathematically section. • Students analyze data to identify linear and nonlinear relationships.
Disciplinary Core Idea(s)	**PS4.A.** Wave Properties • A simple wave has a repeating pattern with a specific wavelength, frequency, and amplitude. • A sound wave needs a medium through which it is transmitted.	• Students construct a wave model to view repeating patterns of waves with varying amplitudes and frequencies. • Students build a model demonstrating wave transfer in a medium. • Students read about how bats use different types of waves.
Crosscutting Concept(s)	• Systems and System Models • Energy and Matter: Flows, Cycles, and Conservation	• Students observe how energy flows from one end to the other in their wave model. • Students observe how energy drives the motion of matter as a wave cycles through matter.
Common Core State Standards (CCSS) Correlations		
Reading Standard(s)	• CCSS.ELA-Literacy.RST.6-8.4. Determine the meaning of symbols, key terms, and other domain-specific words and phrases as they are used in a specific scientific or technical context relevant to grades 6–8 texts and topics.	• The reading skill for this chapter is finding the meaning of new words and keeping up with those meanings throughout a vocabulary-rich article.
Writing Standard(s)	• CCSS.ELA-Literacy.WHST.6-8.2. Write informative/explanatory texts, including the narration of historical events, scientific procedures/experiments, or technical processes. • CCSS.ELA-Literacy.WHST.6-8.9. Draw evidence from informational texts to support analysis, reflection, and research.	• Using their lab observations and the information from an article, students write an explanation to show aspects of waves present in a wave machine.

Chapter 13

Background

This chapter introduces the basic concept of waves by having students build and play with a wave machine. This wave model is excellent at helping students see that the particles in a wave return to their original position rather than flowing with the energy. Students will need to spend some time playing with the wave machine to be ready for the extensive vocabulary words used to describe waves, including *transverse, longitudinal, amplitude, frequency, wavelength,* and *hertz*. Even after using this chapter, you may find it useful to let students revisit this wave machine to help them talk about other topics related to wave motion.

Pre-Reading/Exploration

Materials for Activity (Per Group)

- Safety glasses with side shields or safety goggles
- 2 strips of wide masking tape, each 2 m in length
- 2 ring stands with rings and C-clamps (If you do not have ring stands or clamps, the machines can be attached to any heavy, adjustable objects such as chairs or even student volunteers.)
- Metric ruler or meterstick
- 40 wooden skewers
- 80 gummy candies

> **SAFETY NOTES**
> The following safety recommendations apply to all activities in this chapter:
> - Wear safety glasses with side shields or safety goggles during the setup, hands-on, and takedown segments of the activities.
> - Appropriately dispose of lab materials at the end of the activities as directed by the teacher.
> - Use caution in working with sharp objects (e.g., skewers), which can cut, impale, or scratch skin.
> - Never eat any food items used in a lab activity.
> - Immediately report any lab accident to the teacher.
> - Wash your hands with soap and water after completing the activities.

Activity

Use With Student Page(s): Sweet, Sweet Waves (lab sheet)

In this activity, students will construct a wave model. (See Figures 13.1 and 13.2 on p. 184.) The model will help them understand that in a wave, the particles don't travel with the energy; rather they are temporarily displaced by it. Students will model amplitude, wavelength, and frequency differences by physically interacting with their model. They will be able to transfer a large packet of energy with a single large wave or by sending multiple waves that can add up to the same energy.

This exploration works best if small groups of 6 to 8 students each have their own wave machine to play with. However, using one wave machine as a demonstration is also an option. Students will be most invested in the wave machine if they help build it, so building instructions are written for students. In the instructions, groups of 6 to 8 students work together to build the machine, with 1 to 2 students completing each building task. You

Once Upon a Physical Science Book

All About Bat Waves

Figure 13.1. Front of Wave Machine

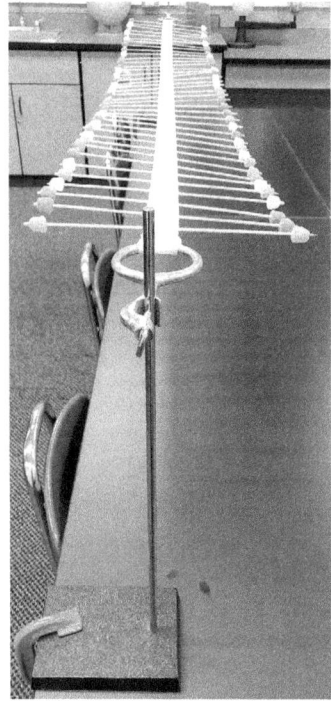

Figure 13.2. Side of Wave Machine

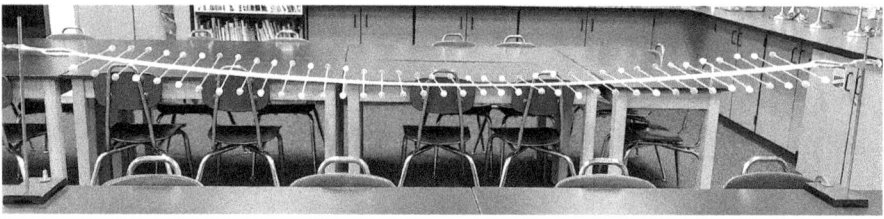

can, however, prepare the machines ahead of time, and even store them for future classes.

Reading

Use With Student Page(s): "Bats: The Night Navigators" (article)

Introduce the Reading. Tell students that they are going to read more about waves in the article "Bats: The Night Navigators." They will be reading about different types of waves, but all waves work like the wave machine, moving energy without sending the particles along.

Reading Strategy: Finding the Meaning of New Words

To introduce the strategy, tell students that scientists use a lot of vocabulary to describe waves and that this article will introduce a lot of new words. It can be especially hard to read text that has lots of new words, but the nice thing about science writing is that it usually tells you what the new words mean. You just have to recognize the definitions.

Display Table 13.1 for students. Have them read through the table. Then ask if they can think of other ways a definition might be given and add any new ideas to the chart. Point out that definitions are usually given just *before* or *after* the *first* use of a new word. If students are looking for a definition and can't find it, they should look back through what they have read to see where the word was introduced.

The first time students read a new word, they will learn its definition. But then the word usually continues appearing in the text. Students should remind themselves of the word's meaning each time they see it. They can mark a word's definition as they read it, making it easy to glance back and remind themselves of the meaning as the word continues to be used. This will help them keep reading as new words accumulate through the passage.

Table 13.1. How to Recognize Word Definitions

Example	Explanation
A bat's vocal cords produce vibrations. Vibrations are quick shaking motions that move back and forth.	The sentence after the term provides a definition.
A bat's vocal cords make quick shaking motions that move back and forth, called vibrations.	The new term is signaled with the word *called*.
A bat's vocal cords produce vibrations, which are quick shaking motions …	The definition is signaled with the phrase *which are* or *which means*.
A bat's vocal cords produce vibrations, or quick shaking motions that move back and forth.	The word *or* after a comma indicates that the vocabulary word and the term or phrase that follows *or* mean the same thing. (This is a signal that many students miss!)
A bat's vocal cords produce vibrations. These quick shaking motions …	This is the trickiest situation. The text doesn't directly say what the word means but implies it by using the word and definition close together.

Journal Question

After students have completed the reading, give them the following question for their reading journals, which will help them internalize the strategy they practiced: When a passage has lots of new words, it can be hard to remember what they all mean. What could you do while reading to help you remember the definitions of all the new words you are learning?

Application/Post-Reading/Writing

- **Writing Prompt.** Think about your wave model. Write a letter to a friend who missed class on the day you used the model. Define the terms *wave, frequency, amplitude,* and *transverse wave* and explain how your friend could see those words in action in your wave model.
  - **Pre-Writing Suggestions.** Students may need to spend a few minutes playing with their wave models while consulting the article to figure out how these four terms relate to the model. Have students jot down notes on what they figure out. Tell your students whether this letter needs to use formal English or if you will accept informal English since it is for a classmate. Because students will be introducing new words in their text, show them Table 13.1 again and encourage them to signal their definitions using the different methods from the table.
  - **Key Evaluation Point.** Students should say that the wave moves energy without moving the particles along, as can be seen by

All About Bat Waves

the sticks returning to their place after the wave passes by. The model shows a transverse wave because the candies bob in a different direction from that of the wave. In the case of this model, frequency describes how often a given candy bobs up and down, and amplitude describes how far each candy bobs up and down.
- **Thinking Mathematically.** Students will use diagrams and measurements in the worksheet A Stormy Sea (p. 193) to observe how changing the amplitude and frequency affect the total energy carried by a wave.

Chapter 13

Sweet, Sweet Waves: Build Your Wave Machine

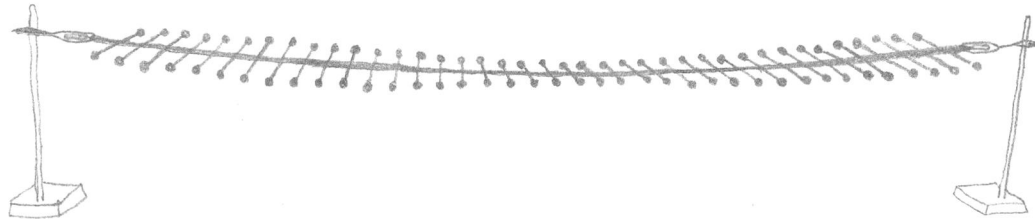

1. After getting the activity materials from your teacher, divide your group into four teams. Each team should complete one of the following tasks:
   - Task 1: Mark the middles of each of the skewers. (Find the middle point by dividing the skewer's length by two.)
   - Task 2: Attach gummy candies to each end of the 40 skewers. Make sure that the candies are all about the same distance from the tips of the skewers.
   - Task 3: Measure a 2 m length of tape. Make a mark every 5 cm along the tape. Allow 10 cm on each end before your first mark.
   - Task 4: Attach a ring stand firmly to the table with a C-clamp. Loosely attach the second ring stand to the table 2 m from the first one. You will adjust this one to put tension on your finished model.

2. Attach each end of the tape to one of the rings. (Wrap each tape end around the ring and back on itself. Make sure not to cover up any of your 5 cm marks.)

3. Lay the skewers with candies across the tape, aligning the marked centers with the middle of the tape. The skewers should be balanced; however, if one or more tips over to one side, they can be adjusted later.

4. Once all the skewers are placed, add the second strip of tape on top of the skewers to hold everything in place. To do this, start by wrapping one tape end over one ring. Work slowly with a partner to match the two pieces of tape so that the sticky sides align.

5. Check if any of your skewers are leaning to one side. If they are, slide the gummy candy on the side that is low toward the center until the skewer balances out. Work from one end of the tape to the other adjusting each candy as needed. Your wave machine should look like a flat bridge when finished.

6. Adjust the tension if your machine sags. Add tension by shifting the looser ring stand back. Less sagging is better. The middle of your bridge shouldn't be more than 10 cm lower than the ends.

Play With Your Wave Machine

1. Walk to one end of the machine and tap the side of a skewer. What happens?

All About Bat Waves

2. Let the machine settle so that it stops moving. Are any of the parts of the machine in a different place than they were when you started?

3. The motion that you saw move along the machine is called a wave. Spend a few moments tapping different skewers to see what kinds of waves you can make. Write three observations from your tapping.

   a. _____

   b. _____

   c. _____

4. Use your hands to steady the skewers so the machine is still. Tap the side of a skewer at the far end of the machine. In which direction did your wave travel?

5. Watch the candies carefully. In which direction do the candies move? Are they moving in the same direction as the wave?

6. The individual candies don't move very far, at least compared to the distance that the wave travels. The tape doesn't move much at all. And all the parts of your wave machine return to the place where they started. That's because the wave carries energy, not particles. Based on this information, try to write a definition for the word *wave*:

7. Start a new wave. Then make a second wave chase the first. Does the second wave ever catch the first wave?

8. Experiment until you figure out how to make bigger and smaller waves. How do you do it?

9. Each time you send a wave down the machine, you are moving energy from one end to another. Which sends more energy, a big wave or a small wave?

10. An individual wave can only get so big. Given that each wave is limited in size, how can you send more energy across the machine?

All About Bat Waves

Bats: The Night Navigators

As dusk falls, a little brown bat pokes its head from its day roost under the eaves of an old barn. It unfolds its wings and flaps them steadily, heading for a pond where mosquitoes swarm.

A bat on the hunt is a mysterious sight, if you can see it at all. On this night, like many nights, the little brown bat hunts in near-total darkness. Early bat scientists were baffled by bat navigation—in their experiments, even blindfolded bats could fly without crashing! How did the animals find their way around obstacles and catch insects when they couldn't see? It took almost 200 years for scientists to figure out that bats navigate using sound.

> REMEMBER YOUR CODES
> ! This is important.
> ✓ I knew that.
> X This is different from what I thought.
> ? I don't understand.

Sound Makes Waves

To make a sound, the bat vibrates vocal cords in its throat. Each vibration gives the air nearby a shove. That shove pushes those air molecules into the neighboring molecules, which then knock into the next bunch of molecules, and so on, as shown in Figure S13.1. Individual molecules only move a little as they knock back and forth, and each molecule ends up back where it started. But the energy keeps moving forward through the air. This creates a wave, or a disturbance that moves energy from one place to another.

In sound waves, the molecules are knocked back and forth along the same line in which the energy is traveling. Waves in which the molecules and energy travel in the same plane are called longitudinal waves. But not all waves follow this pattern.

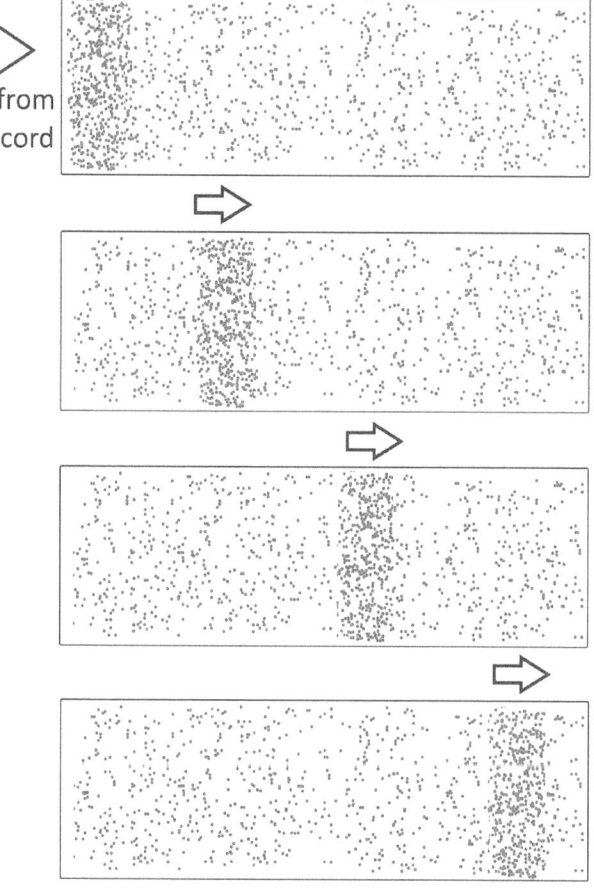

Figure S13.1. A Sound Wave Compresses Air

Push from vocal cord

Chapter 13

Another Type of Wave

Picture the little brown bat's wing flapping up and down. If you traced the up-and-down path taken by the tip of the wing, you would see another type of wave generated by the bat called a transverse wave. In a transverse wave, the energy and the molecules move in different directions as shown in Figure S13.2. For example, if a transverse wave were moving horizontally across the room, the molecules in the wave would be bobbing up and down.

Notice that the transverse wave has peaks and dips. Scientists use those peaks and dips to describe the wave. If you measure the distance from the top of one peak to the top of the next peak, you have the wavelength, as shown in Figure S13.3. You can also count how many peaks move past a certain spot within a certain amount of time. That number is called the frequency because it tells you how *frequently* waves pass by. Frequency is measured in Hertz (Hz), which is the number of peaks that pass by in one second.

Figure S13.2. Movement in Longitudinal and Transverse Waves

Longitudinal Wave:

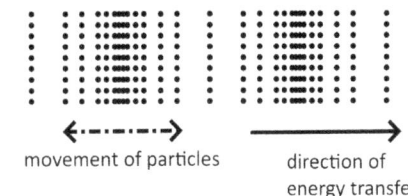

Transverse Wave:

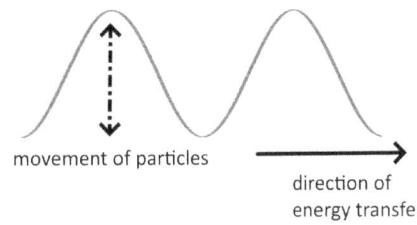

Figure S13.3. High-Frequency and Low-Frequency Wave

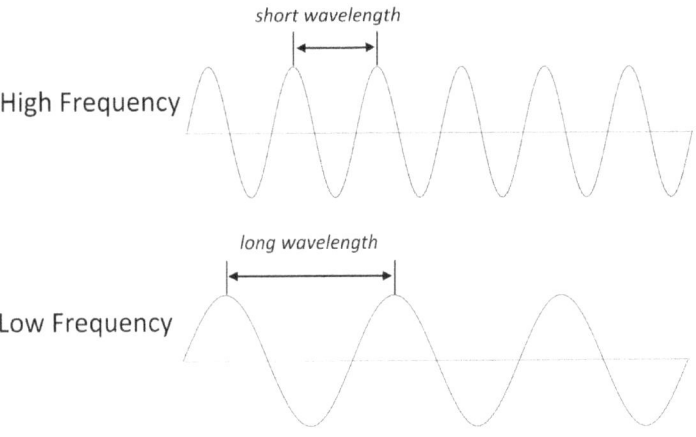

Scientists have similar methods to describe longitudinal waves. Look back at Figure S13.2. Notice how sections of compressed particles alternate with sections where the particles are more spread out. The compressed sections are like the peaks on a transverse wave. They are used to measure the wavelength and frequency of longitudinal waves.

Seeing With Echoes

How do bats use waves to find their way in the dark? Sound waves can reflect, or bounce off, surfaces. The reflection of a sound is called an echo. When longitudinal waves bump a bat's eardrum, the bat hears a sound. Slow, low-frequency waves have a deep sound whereas quick, high-frequency waves are high-pitched. Bats listen for the echoes of their calls to figure out the location of objects. Therefore, bat navigation is called echolocation.

All About Bat Waves

The call of a little brown bat is definitely high-pitched. It's so high, in fact, that humans cannot hear it. The human ear can capture waves with frequencies of about 20 to 20,000 Hz. Most bats, however, echolocate at frequencies of 20,000 to 200,000 Hz. That's 200,000 compressions of air in a single second! These waves create sounds that are too high for our ears to detect.

Amped-Up Amplitude

Those early scientists who were blindfolding bats did not know the bats were using sound to find their way, because the sounds were too high for the scientists to hear. That's a good thing because bat screams are LOUD. Loud sound waves carry lots of energy. The amount of energy carried by a wave is called its amplitude. When graphing a longitudinal wave, the amplitude is a measure of how compressed the air gets. On a graph of a transverse wave, the amplitude is the distance from the midpoint to the top of a peak (as shown in Figure S13.4.) The call of the little brown bat carries so much energy that it can damage the bat's own ears! Fortunately, bats have muscles in their ears that help protect them from the force of their echoes.

As the little brown bat heads for the pond, it vibrates its vocal cords to send out a steady stream of sound waves. They come fast and furious, with short wavelengths, high frequencies, and high amplitudes. When echoes bounce back to its ears, the bat swoops and swerves to avoid tree trunks and branches. Finally, it is rewarded with mouthfuls of juicy mosquitoes.

The Big Question

Draw two transverse waves: (a) one with a long wavelength and high amplitude, and (b) one with a short wavelength and low amplitude.

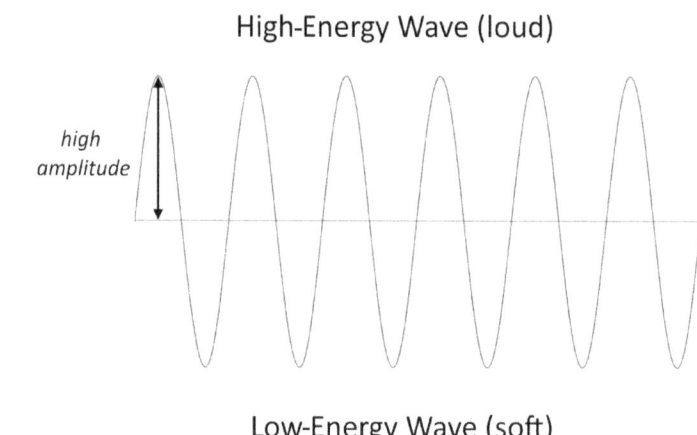

Figure S13.4. High- and Low-Energy Waves

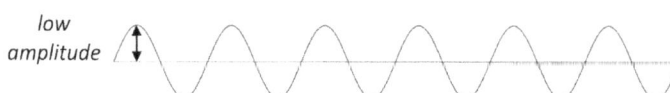

Chapter 13

Thinking Mathematically: A Stormy Sea

Picture a boat bouncing up and down on the waves, as shown in Figure S13.5. Now imagine the graph in Figure S13.6 shows movement of the waves over time.

Figure S13.5. Boat on Waves

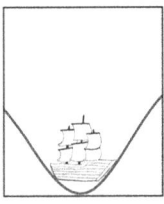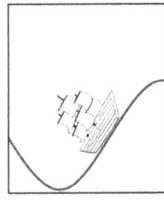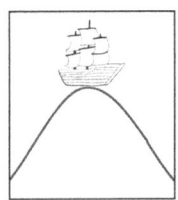

Figure S13.6. Movement of Waves Over Time

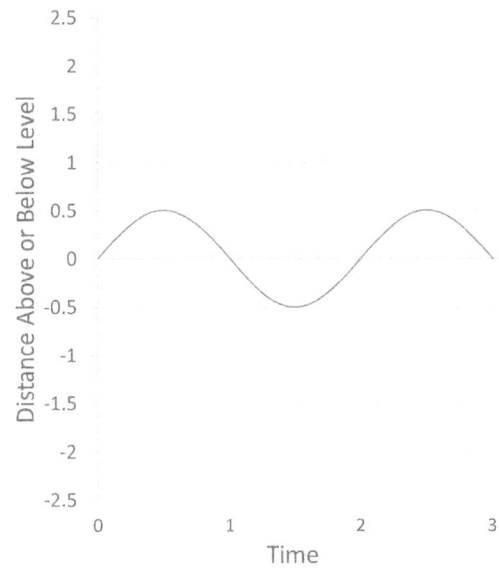

Say that your boat didn't float very well and each time a wave came by, it was *your* job to lift the boat to the top of the wave. Figure S13.7 shows what the waves were like on Monday and Figure S13.8 shows what the waves were like on Tuesday:

Figure S13.7. Waves on Monday

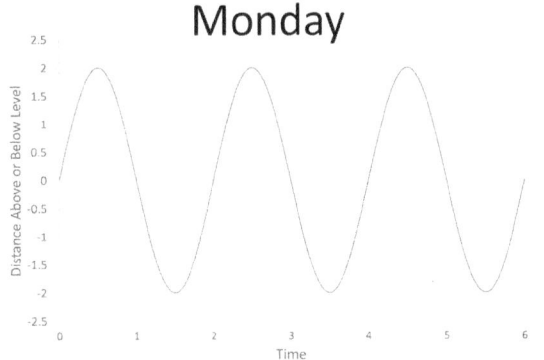

Figure S13.8. Waves on Tuesday

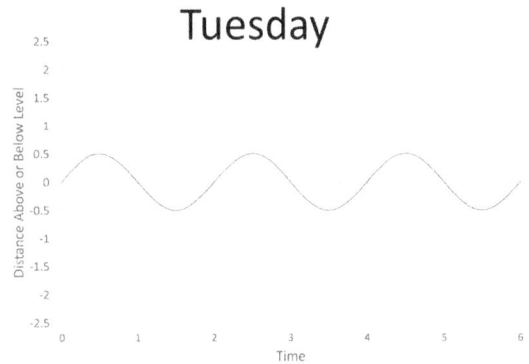

1. On which day would you need the most energy to do the lifting?

On Monday and Tuesday, the frequency of the waves was the same, but the amplitude was higher on Monday. There is a set mathematical relationship between the amplitude of a wave and the amount of energy it carries. Consider the information displayed in the following chart about a certain wave:

Once Upon a Physical Science Book

193

All About Bat Waves

Amplitude (cm)	Energy in the Wave (joules)
1	1
2	4
3	9
4	16
5	25

2. Based on this chart, when you increase the amplitude of a wave, the energy (circle one) increases/decreases.

3. Does it increase or decrease by the same amount each time?

To find the mathematical relationship between amplitude and energy, it may be helpful to fill in the chart that follows. To fill in the blanks, figure out what you need to multiply the amplitude by to get the energy:

Amplitude (cm)	Multiply by?	Energy (joules)
1	× _____	= 1
2	× _____	= 4
3	× _____	= 9
4	× _____	= 16
5	× _____	= 25

4. What is the relationship between amplitude and the energy carried in a wave?

On Wednesday and Thursday, the amplitude of the waves stayed the same, but a strong wind came through on Thursday. Here are the waves from Wednesday (Figure S13.9) and Thursday (Figure S13.10):

Figure S13.9. Waves on Wednesday

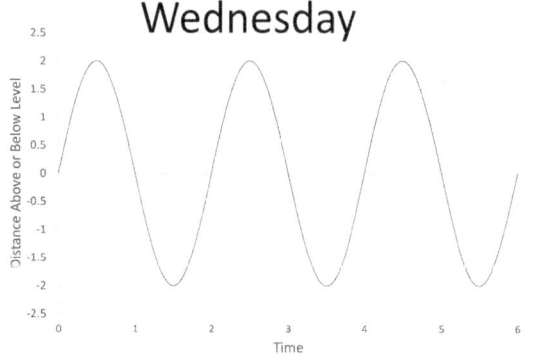

Figure S13.10. Waves on Thursday

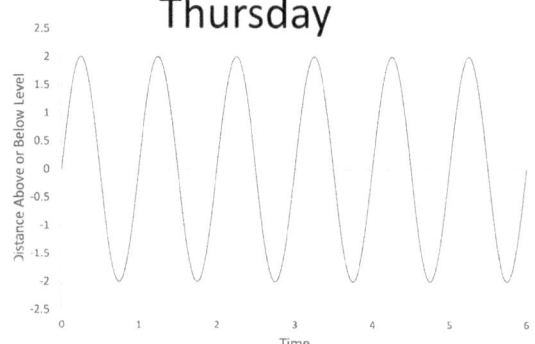

5. On which day would you need the most energy to do the lifting? _____

There is a set mathematical relationship between the frequency of a wave and the amount of energy it carries. Consider the information in the following chart about a certain set of waves:

Frequency (Hz)	Energy in the Wave (joules)
1	1
2	2
3	3
4	4
5	5

6. Based on this chart, what happens to the energy carried in the waves when you increase the frequency by 1 Hz?

7. If you increase the frequency by 2 Hz, what happens to the energy?

8. In general, if you increase the frequency of any wave, you also increase its _____.

9. You can see that there are two ways to change the energy carried in a wave. The first is by changing the _____. The second is by changing the _____.

10. Which creates a greater change in energy: changing the frequency or changing the amplitude?

Chapter 14
Lighting the Way

Topics
- Waves in communication
- Digital and analog systems
- Fiber optics

Reading Strategy
- Talk your way through it

Connections to Standards

Next Generation Science Standards (NGSS) Correlations	
Standard **MS-PS4.** Waves and Their Applications in Technologies for Information Transfer (*www.nextgenscience.org/dci-arrangement/ms-ps4-waves-and-their-applications-technologies-information-transfer*)	
Performance Expectation(s) The materials/lessons/activities outlined in this chapter are just one step toward reaching the performance expectation(s) listed below. **MS-PS4-3.** Integrate qualitative scientific and technical information to support the claim that digitized signals are a more reliable way to encode and transmit information than analog signals.	

Dimension	Element	Matching Student Task or Question From the Activity
Science and Engineering Practice(s)	• Obtaining, Evaluating, and Communicating Information	• Students send and receive short codes using analog and digital systems. • Students practice using binary code to see how computers transmit large amounts of information in the Thinking Mathematically section.
Disciplinary Core Idea(s)	PS4.C. Information Technologies and Instrumentation • Digitized signals (sent as wave pulses) are a more reliable way to encode and transmit information.	• Students send and receive short codes using analog and digital systems, discovering that the digital system is more accurate. • Students read about how analog and digital systems are used to transmit sound over long distances.
Crosscutting Concept(s)	• Structure and Function	• Students send and receive short codes using analog and digital systems, visualizing the relationship among the parts of each system. • Students read about how the refracting structure of fiber optics functions to send sound at a fast, reliable rate.

Common Core State Standards (CCSS) Correlations	
Reading Standard(s)	• CCSS.ELA-Literacy.RST.6-8.2. Determine the central ideas or conclusions of a text; provide an accurate summary of the text distinct from prior knowledge or opinions.	• The reading skill for this chapter is talk your way through it, which is a summarizing and remembering strategy for complex text.
Writing Standard(s)	• CCSS.ELA-Literacy.WHST.6-8.2. Write informative/explanatory texts, including the narration of historical events, scientific procedures/experiments, or technical processes. • CCSS.ELA-Literacy.WHST.6-8.2.c. Use appropriate and varied transitions to create cohesion and clarify the relationships among ideas and concepts.	• Students write an explanation for how sound travels through different media between two friends who are having a computer chat. Students are specifically asked to brainstorm transition words that can help their writing express chronology.

Chapter 14

Background

Tying together waves and communication, this chapter works well after basic information on waves, reflection, and refraction have been taught. It also addresses digital versus analog communication, which is included in the *NGSS*.

The chapter's main exploration has students trying to send and receive short codes using an analog system and a digital system. Students will quickly determine that the digital system is more accurate. Then they will read about how a combination of analog and digital systems are used in modern communication systems.

Pre-Reading/Exploration

Materials for Activity

- Copies of digital and analog codes and transmission sheets for each group
- Slips of paper with five-digit codes of your choosing
- Pencils
- Inexpensive toy with fiber optic strands (optional)

Activity

Use With Student Page(s): Two Codes (student handout), Code Transmission 1: Analog (student handout), and Code Transmission 2: Digital (student handout)

Ask students to think through the types of waves you have talked about and propose which ones might be used for communication. Radio waves and sound waves will probably come up, as may some others. Tells students that you are going to be exploring how waves carry information, and you are going to start by playing a game with codes.

Divide students into teams of 4 to 6 players. Each team should then divide into a "code-sending group" and a "code-receiving group." Have the code-sending players go to one side of the room and the code-receiving players go to the other. Place a table or basket in the center for students as the transmission point where the sending team can leave the code for the receiving team. (See Figure 14.1, p. 200.)

Figure 14.1. Setup for Sending Codes

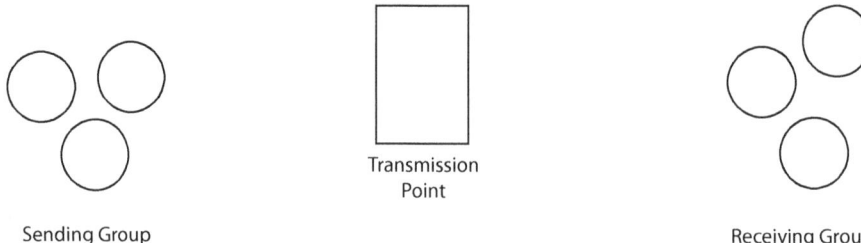

Give each sending group and each receiving group a copy of the analog code (p. 203) and several pencils. (*Note*: At this point, it does not matter if students attach any meaning to the word *analog* other than that it is the name of the first code.)

Give each sending group the Code Transmission 1 sheet (p. 204), along with a secret five-digit code of your choosing. You can give each team the same code, or vary them. They should color in the blocks on their transmission sheet and then run the code to the transmission point. Their receiving group will grab it and translate it using their own copy of the code. This first round is a practice round. Students may complain that copying and interpreting the colors is difficult. Tell them to do the best they can. (After all, the difficulty of analog coding is what we ultimately want students to understand.) Once all the teams have practiced sending a message, have a race. The first team to send their code and have it interpreted correctly wins.

Next, give copies of the digital code (p. 203) to each code-sending and code-receiving team. Give the sending teams copies of the Code Transmission 2 sheet (p. 205). Repeat the process you completed for the analog code, letting teams try to transmit a secret digital code one time for practice and then in a race.

Tell students that for the final round, they can select which type of code to use to send their message. Sending and receiving groups should meet together. They must (1) decide whether their team wants to use the analog or digital code for the last race and (2) be ready to say why they selected the code they chose. Before the final race, have each group quickly share which code they want to use and why. Most will choose digital, but whatever they select is fine. Provide the sending team with a new secret code and hold the final race.

Before students complete the reading, it may be helpful to show them a toy made with fiber optics. Such toys are available from novelty companies at low cost. The fiber optics on these toys are not perfect—they are designed to let some light escape as it travels. In contrast, fiber optic cables used for communication do not appear to be lit at all except at the tips. Nevertheless, the toys will give students a mental image that will help them understand how fiber optics work.

Reading

Use With Student Page(s): "The Sweet Sound of Light"

Introduce the Reading. Tell students they are going to read about how analog and digital codes are used in everyday communications.

Reading Strategy: Talk Your Way Through It

Project the following excerpt from the reading and read it aloud:

> *Microphones take precise measurements of the changes in pressure generated by sound waves. They detect each of the components of waves that you have learned about: amplitude, wavelength, speed and frequency, plus other information, such as the direction the sound is coming from.*
>
> *Each of these measurements falls on a continuum. The frequency of an audible sound wave, for example, could fall anywhere from 20 to 200,000 Hz. It is as if the microphone is looking at blocks colored in a range of gray shades and recognizing each tiny change of color (Figure S14.1). This type of information—data that can fall continuously over a broad range—is called analog.*

The talk your way through it strategy was first covered in Chapter 9. If you haven't already used this chapter with your class, read pages 122–123 for information on how to introduce the strategy to students. If you have already introduced the strategy, point out that this excerpt contains a lot of information. Ask students to suggest a strategy that would help them keep up with what they are learning as they read. Accept any reasonable suggestion, but turn their attention to the talk your way through it strategy as a possibility.

Ask a student to model what they might say to themselves after reading the passage. Then ask other students for information they might add to the first student's summary. Ask if there are specific ideas they think they might jot down in the margin to help them remember if they need to look back and find information as they read. You might suggest that they underline the word *analog* so that if it shows up later, they will be able to quickly remind themselves what it means.

Tell students that other paragraphs in this article are also dense and may contain information that students have not learned before. Instruct them to try pausing after each section to talk through (out loud, silently, or on paper) what they read before moving on. If students are using reading groups, have them spend extra time filling in the details when they reach the part of the group-reading process in which one group member describes what they read.

Journal Question

After students have completed the reading, give them the following question for their reading journals, which will help them internalize the strategy they practiced: Do you already use a strategy like talk your way through it when you read books with lots of information? If so, describe how you do it. If not, describe how the talk your way through it strategy works so you can try it in the future.

Application/Post-Reading/Writing

- **Writing Prompt.** Imagine that you are talking to a friend over a video chat. Trace the path of the message from your mouth to your friend's ears. At each step, explain what type of waves are involved and whether the message is analog or digital.
  - **Pre-Writing Suggestions.** You may need to help students think about the message transmission process by drawing a diagram of it before they write. Ask students what science words they will want to include in their responses. Also point out that this writing assignment features a sequence and ask what are some words that the students could use to describe the sequence. (Answer: *first, second, then, finally.*)
  - **Key Evaluation Point.** Students should say that the message begins in sound waves, which are analog, and is converted to digital inside the microphone or computer. From there, the message stays digital but may follow a variety of routes. Students may discuss sending the message through electrical wires, or it may go wireless via radio waves. Alternatively, it may go through optical cables (infrared light waves). In the receiving computer and speaker, the information is turned back into analog sound waves and travels through the air to the listener.
- **Thinking Mathematically.** With the worksheet The Power of Two (p. 209), students are introduced to making digits using base 2, which is the basis of digital code.

FIND OUT MORE
For a demonstration on how sound waves can be carried through wires or as a light wave, visit *www.exploratorium.edu/snacks/modulated-led.*

Chapter 14

Two Codes

Code 1: Analog

Code 2: Digital

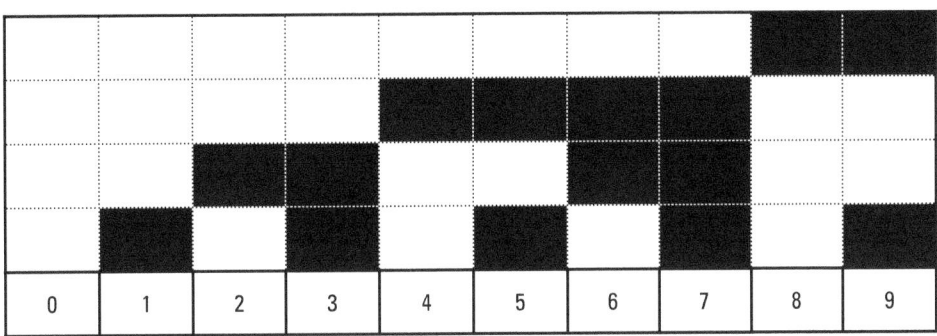

Once Upon a Physical Science Book

Code Transmission 1: Analog

Practice Round:

Sending team, put code here:

Receiving team, translate code here: _____ _____ _____ _____ _____

Race Round:

Sending team, put code here:

Receiving team, translate code here: _____ _____ _____ _____ _____

Extra Round:

Sending team, put code here:

Receiving team, translate code here: _____ _____ _____ _____ _____

Chapter 14

Code Transmission 2: Digital

Practice Round:

Sending team, put code here:

Receiving team, translate code here: _____ _____ _____ _____ _____

Race Round:

Sending team, put code here:

Receiving team, translate code here: _____ _____ _____ _____ _____

Extra Round:

Sending team, put code here:

Receiving team, translate code here: _____ _____ _____ _____ _____

Lighting the Way

The Sweet Sound of Light

Your favorite band is touring, but the closest concert is 300 miles away. You make do with the next best thing. The concert is streaming live, and you can listen on your phone.

Far away at the concert, the guitar starts up. It generates sound waves, or energy patterns that move through molecules in the air. How can a pattern in air molecules 300 miles away travel all the way to your phone to make sounds?

> **REMEMBER YOUR CODES**
> ! This is important.
> ✓ I knew that.
> X This is different from what I thought.
> ? I don't understand.

The first step in the journey involves a microphone (or, more likely, several microphones). Microphones take precise measurements of the changes in pressure generated by sound waves. They detect each of the components of waves that you have learned about: amplitude, wavelength, speed and frequency, plus other information, such as the direction the sound is coming from.

Each of these measurements falls on a continuum. The frequency of an audible sound wave, for example, could fall anywhere from 20 to 200,000 Hz. It is as if the microphone is looking at blocks colored in a range of gray shades and recognizing each tiny change of color (Figure S14.1). This type of information—data that can fall continuously over a broad range—is called analog.

Figure S14.1. Gray Shades

Analog information falls along a range, such as dark to light or 1 to 100.

Analog or Digital?

A microphone collects analog information about the music in the form of measurements. Then a computer takes each of those measurements and converts it into a sort of code. In this digital code, the information can be shared using just two values. The two values could be one and zero, black and white, on and off, and so on. (See Figure S14.2.)

Figure S14.2. One Type of Digital Code

Digital information has just two values, such as black and white or one and zero.

It may seem extremely complicated to turn so many measurements into ones and zeros. And it is. To illustrate how it's done, let's return to the idea of using two colors—black and white—to code information. Say that you have a piece of information that is a particular shade of gray and you want to translate it into code using

Figure S14.3. An Analog Shade of Gray Coded as a Black-and-White Pattern

the black-and-white coding system. The code might turn out similar to the one in Figure S14.3.

It may be more complicated to create a digital code than an analog code, but it is much easier to *send* digital information, as it only has two values. You can use electricity, using a zap of energy to represent the one and a break in the energy to represent the zero. You can use sound—a long beep for one, a short beep for zero. You can use light, turning the light on for one and off for zero. Transmitting a code with only two values cuts down on misunderstandings. Think back to the analogy of translating a piece of information that is shaded gray into a black-and-white code. Now say this code is sent out. The receiver will still be able to read and understand the code, even if there are small errors—say, the black isn't perfectly black or the white isn't perfectly white. This is much more efficient than sending out information shaded in gray where the receiver has to figure out the exact shade in order to interpret the code.

Once a computer has converted a message to digital code, the code can be sent through wires. Back at the concert, the information about the music may start out traveling through electric wires. But as the information begins the long journey to your phone, it will probably move as bursts of light through fiber optic cables.

Fiber Optics

Light naturally spreads out, so fiber optic cables have a special design to keep the waves moving in the right direction. These cables are made of dense glass or plastic surrounded by a layer of less-dense glass or plastic (Figure S14.4). When light waves move from a dense material into a less-dense material, the light bends through a process called refraction. If the angle of the bend is steep enough, all the light refracts right back into the material it came from.

Figure S14.4. How Light Moves Through a Fiber Optic Cable

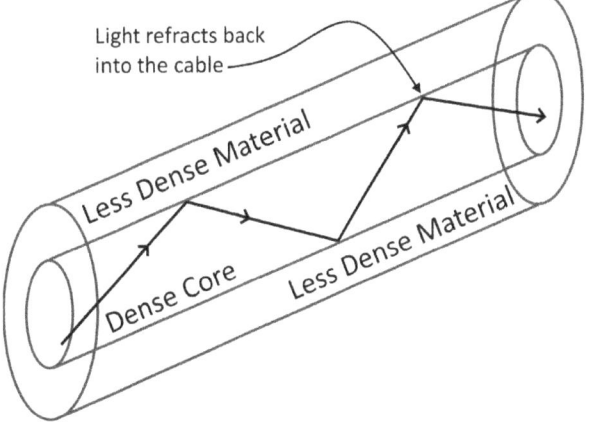

Lighting the Way

Light waves, usually of infrared light, travel through the dense material in the center of the cable. If the light hits the less-dense side, it bounces back. Light moves faster than anything else, so fiber optics create a fast, reliable way to send information over long distances.

Streaming to Your Phone

The digital information describing your band's opening song zips along miles of fiber optic cable, encoded in light waves. Eventually, that information reaches the cell towers communicating with your phone. The cell towers use radio waves to keep the message heading your way. Radio waves have longer wavelengths than the infrared waves used in fiber optic cables and can travel through walls and roofs.

Mere moments after the band cranks up, opening chords stream from the speakers in your phone. The digital message has been returned to analog. The computer in your phone reads the digital code that described the original sound waves. Moving parts in your phone's speaker pulse against the air molecules to re-create those waves. The sound waves pass through the air until they reach your ear drum, and soon you are singing and dancing around your room.

The Big Question

Why is digital information more likely to be accurate than analog information?

Chapter 14

Thinking Mathematically: The Power of Two

Since digital codes only use two values, they are also called binary codes. (The word *binary* means something made of two parts.) Computers use a binary code made of ones and zeros. How do computers use only two digits to represent larger numbers?

First, it helps to understand our standard numbering system. In our usual system, each digit tells how many ones, tens, or multiples of ten we have:

- In the above image, how many "thousands" are there? _____
- How many "hundreds" are there? _____
- How many "tens" are there? _____
- How many "ones" are there? _____

In binary code, each digit tells us how many ones, twos, or multiples of two we have:

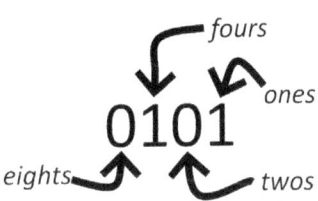

You can imagine it as a series of columns that you fill with black dots. Each column must be all the way full or all the way empty:

- In the example on the right, how many ones are there? _____
- How many twos are there? _____
- How many fours are there? _____
- How many eights are there? _____
- How many black dots are there in total? _____

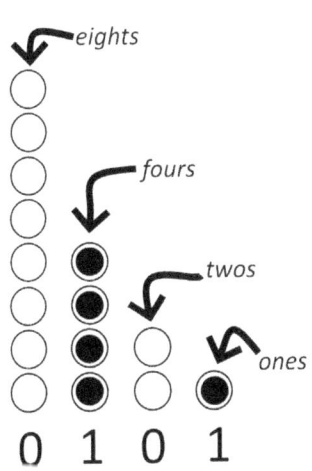

Lighting the Way

As you can see, the number 5 is represented as 0101 in binary code. Figure out how to write the number 6 in binary code, using the following visual. Remember, each column must be all the way full or all the way empty:

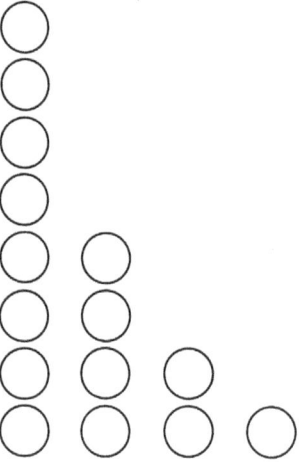

- The number 6 in binary code is: _____

Use the dot system to try to translate each of the following binary codes into numerals:

- 1100 _____
- 1001 _____
- 0011 _____
- 1111 _____ (*Note:* This is the largest number you can write with only four columns. But with more columns, you could write larger numbers.)

Image Credits

Superhero images (pages 1, 11, 23, 43, 59, 77, 91, 105, 119, 135, 151, 165, 181, 197): Shutterstock

Chapter 1

Figure 1.1: Authors

Chapter 3

Figure S3.1: Authors
Figure S3.2: Shutterstock
Figure S3.3: Shutterstock
Figure S3.4: Authors
Figure S3.5: Pixabay, Public domain. https://pixabay.com/illustrations/car-accident-car-crash-car-accident-1995852.

Chapter 4

Figure S4.1: Authors
Figure S4.2: Authors
Figure S4.3: Authors
Jars: Pixabay, Public domain. https://pixabay.com/vectors/jar-glass-empty-jar-glass-jar-162166.

Chapter 5

Figure 5.1: Authors
Figure 5.2: Authors
Figure S5.1: Pixabay, Public domain. https://pixabay.com/photos/husky-dog-face-eyes-fur-animal-274611.
Chemical reactions (pp. 71–76): Authors

Chapter 6

Tape dispenser: Shutterstock
Tape strip: Authors
Table: Pixabay, Public domain. https://pixabay.com/vectors/table-furniture-wooden-1300555.
Figure S6.1: Shutterstock
Figure S6.2, sweatshirt: Pixabay, Public domain. https://pixabay.com/illustrations/fashion-clothing-sweater-shirt-902657.
Figure S6.2, sock: Pixabay, Public domain. https://pixabay.com/vectors/sock-foot-feet-wool-clothing-145589.
Figure S6.3: Shutterstock
Figure S6.4: Authors

Image Credits

Chapter 7

All images in chapter by authors

Chapter 8

Figure 8.1: Authors
Figure 8.2: Authors
Figure 8.3: Authors
Figure S8.1: Authors
Figure S8.2: Authors
Figure S8.3: Shutterstock
Figure S8.4: Shutterstock

Chapter 9

Figure S9.1: Authors
Figure S9.2: Authors
Figure S9.3: Authors
Figure S9.4: Authors
Figure S9.5: Authors
Figure S9.6: Pixabay, Public domain. https://pixabay.com/photos/roller-coaster-ride-amusement-park-3619387.
Figure S9.7: Pixabay, Public domain. https://pixabay.com/vectors/boat-navy-ocean-sea-submarine-2026130.
Thinking Mathematically roller coaster diagram: Authors

Chapter 10

Figure S10.1: Wellcome Collection, CC BY 4.0. https://wellcomecollection.org/works/tn8y3dfr?query=L0023106.
Figure S10.2: Shutterstock
Figure S10.3: Pixabay, Public domain. https://pixabay.com/illustrations/thermometer-temperature-measure-1917500.
Figure S10.4: Authors

Chapter 11

Figure 11.1: Authors
Figure 11.2: Authors
Figure S11.1: Jay Rotella, Weddell Seal Project, photo of Weddell seal was obtained under NMFS Permit 21158.
Figure S11.2: Pixabay, Public domain. https://pixabay.com/vectors/balding-man-face-brown-head-31110.
Figure S11.3: Pixabay, Public domain. https://pixabay.com/photos/feather-down-feather-bird-feather-648481.
Figure S11.4: Jay Rotella, Weddell Seal Project
Figure S11.5: Shutterstock

Chapter 12

Figure S12.1: Authors
Figure S12.2: Authors
Figure S12.3: Authors
Figure S12.4: Authors
Figure S12.5: Authors
Figure S12.6: Shutterstock
Figure S12.7: Shutterstock
Figure S12.8: Authors

Chapter 13

All images in chapter by authors

Chapter 14

All images in chapter by authors

Index

Note: Page numbers in **boldface** type refer to tables or figures.

A

"All About Bat Waves," 181–195
   background, 183
   "Bats: The Night Navigators" reading, 184, **185, 190,** 190–192, **191, 192**
   connections to standards, 182
   journal questions, 185
   pre-reading/exploration, 183
   Sweet, Sweet Waves activity, 183–184, **184,** 187–189
   Thinking Mathematically: A Stormy Sea, 186, 193–195
"All Charged Up," 77–89
   application/post-reading/writing, 81
   background, 79
   connections to standards, 78
   journal questions, 81
   "One Clingy Sock" reading, 80, 85–87
   pre-reading/exploration, 79
   Stuck on Sticky Tape activity, 79–80, 82–84
   Thinking Mathematically: Electric Field Forces, 81, 88–89
animal kicks. *See* "Kick Force"
application phase of learning cycle. *See* concept application phase of learning cycle
assessment of student learning, 6–7, **7, 8**
   self-assessment of strategy use, 19–20, **20**

B

bat echolocation. *See* "All About Bat Waves"
Big Question, 6
The BSCE 5E Instructional Model, 2

Index

C

cause and effect, **18, 139,** 139–140
charges. *See* "All Charged Up"
chemical hand warmers. *See* "Handy Heaters"
chemical properties, 47. *See also* "Identify This"
chemical reactions. *See* "Handy Heaters"
chunking (strategy), 18
   for "All Charged Up," 80
   for "Identify This," 48–49
claims, evidence, and reasoning, 7, **8**
classroom culture, 12
Common Core State Standards (CCSS), 2
   lesson connections to
      for "All About Bat Waves," 182
      for "All Charged Up," 78
      for "Ding-Dong Electromagnets," 166
      for "Energy's Wild Ride," 120
      for "Handy Heaters," 61
      for "How to Not Die in Antarctica," 152
      for "Identify This," 44
      for "Inertia: It's a Drag," 92–93
      for "Kick Force," 106
      for "Lighting the Way," 198
      for "Taking Your Temperature," 136–137
      for "The Smash-Masters," 24
compare and contrast
   for "How to Not Die in Antarctica," 154–155
comparisons, **18**
comprehension coding, 14, 15
   for "The Smash-Masters," 26
concept application phase of learning cycle, 3
concept application phase of the learning cycle
   for "All About Bat Waves," 185–186
   for "All Charged Up," 81
   for "Ding-Dong Electromagnets," 169–170
   for "Energy's Wild Ride," 123–124
   for "Handy Heaters," 66–67
   for "How to Not Die in Antarctica," 156
   for "Identify This," 49, 57
   for "Inertia: It's a Drag," 97
   for "Kick Force," 110–111
   for "Lighting the Way," 202

for "Taking Your Temperature," 140–141
　　　for "The Smash-Masters," 28, 38–42
conservation of energy. See "Energy's Wild Ride"
constructivism, 3
contrasts, **18**
crosscutting concepts
　　　for "All About Bat Waves," 182
　　　for "All Charged Up," 78
　　　for "Ding-Dong Electromagnets," 166
　　　for "Energy's Wild Ride," 120
　　　for "Handy Heaters," 60
　　　for "How to Not Die in Antarctica," 152
　　　for "Identify This," 44
　　　for "Inertia: It's a Drag," 92
　　　for "Kick Force," 106
　　　for "Lighting the Way," 198
　　　for "Taking Your Temperature," 136
　　　for "The Smash-Masters," 24

D

diagrams and illustrations, 17
"Ding-Dong Electromagnets," 165–179
　　　application/post-reading/writing, 169–170
　　　background, 167
　　　Building Strength activity, 169, 176–177
　　　connections to standards, 166
　　　journal questions, 169
　　　pre-reading/exploration, 167
　　　Recycling Quick Sort activity, 167–168, **171**, 171–172, **172**
　　　Thinking Mathematically: Using Data to Make a Prediction, 170, 178–179
　　　"When the Pizza Came" reading, 168–169, **173**, 173–175, **174**, **175**
disciplinary core ideas
　　　for "All About Bat Waves," 182
　　　for "All Charged Up," 78
　　　for "Ding-Dong Electromagnets," 166
　　　for "Energy's Wild Ride," 120
　　　for "Handy Heaters," 60
　　　for "How to Not Die in Antarctica," 152
　　　for "Identify This," 44
　　　for "Inertia: It's a Drag," 92
　　　for "Kick Force," 106

Index

for "Lighting the Way," 198
for "Taking Your Temperature," 136
doorbells. *See* "Ding-Dong Electromagnets"

E

echolocation. *See* "All About Bat Waves"
electric fields. *See* "All Charged Up"
electromagnets. *See* "Ding-Dong Electromagnets"
energy and particle movement. *See* "Taking Your Temperature"
"Energy's Wild Ride," 119–133
    application/post-reading/writing, 123–124
    background, 121
    Building Your Kinetic Kar activity, 121–122, **128,** 128–129, **129**
    connections to standards, 120
    Energy on the Move activity, 121–122, **125,** 125–127
    "Energy's Wild Ride" reading, 122–123, 130–132, **131**
    pre-reading/exploration, 121
    journal questions, 123
    safety notes, 122
    Thinking Mathematically: Up and Down the Gravity Road, 124, 133
energy transformation and efficiency. *See* "Energy's Wild Ride"
explanation phase of learning cycle, 3
exploration phase of learning cycle, 2–3
    for "All About Bat Waves," 183
    for "All Charged Up," 79
    for "Ding-Dong Electromagnets," 167
    for "Energy's Wild Ride," 121
    for "Handy Heaters," 61–62
    for "How to Not Die in Antarctica," 153
    for "Identify This," 45–46
    for "Inertia: It's a Drag," 93
    for "Kick Force," 107
    for "Lighting the Way," 199
    for "Taking Your Temperature," 137
    for "The Smash Masters," 25–27

F

fiber optics. *See* "Lighting the Way"
finding the meaning of new words (strategy), 16
    for "All About Bat Waves," 184, **185**

Index

H

"Handy Heaters," 59–76
   application/post-reading/writing, 66–67
   background, 61
   connections to standards, 60–61
   "Frostbite Free" reading, 65–66, 71–73
   Handy Heaters activity, 62–65, **63, 64,** 68–70
   journal questions, 66
   pre-reading/exploration, 61–62
   Thinking Mathematically: Balancing Chemical Reactions, 67, 74–76

heat. *See* "Taking Your Temperature"

heat transfer. *See* "How to Not Die in Antarctica"

"How to Not Die in Antarctica," 151–163
   background, 153
   connections to standards, 152
   Controlled Experiment: Bundle Up! activity, 153–155, **154,** 157–158
   "How to Not Die in Antarctica" reading, 154–155, **159,** 159–161, **160, 161**
   journal questions, 155
   pre-reading/exploration, 153
   Thinking Mathematically: A Different Kind of Wolf, 156, **162,** 162–163

I

"Identify This," 43–57
   application/post-reading/writing, 49, 57
   background, 45
   connections to standards, 44
   journal questions, 49
   pre-reading/exploration, 45–46
   safety notes, 45, 50
   Same or Different? lab sheet, 46–48, 50–53
   Thinking Mathematically: Powder Confusion, 57
   "Was It a Drug Bust?" reading, 48–49, 54–56

"Inertia: It's a Drag," 91–103
   application/post-reading/writing, 97
   background, 93
   connections to standards, 92–93
   Hold On! activity, 94–95, 98
   journal questions, 97
   Marble Maze activity, 95–96, 99
   materials, 93
   "One Long Bus Ride" reading, 96–97, 100–102

pre-reading/exploration, 93
Thinking Mathematically: Out of Balance?, 97, 103
insulation. *See* "How to Not Die in Antarctica"

J

journal questions, 6
   for "All About Bat Waves," 185
   for "All Charged Up," 81
   for "Ding-Dong Electromagnets," 169
   for "Energy's Wild Ride," 123
   for "Handy Heaters," 66
   for "How to Not Die in Antarctica," 155
   for "Identify This," 49
   for "Inertia: It's a Drag," 97
   for "Kick Force," 110
   for "Lighting the Way," 202
   for "Taking Your Temperature," 140
   for "The Smash-Masters," 28

K

"Kick Force," 105–118
   application/post-reading/writing, 110–111
   background, 107
   connections to standards, 106
   journal questions, 110
   "Of Kangaroos and Secretary Birds" reading, 109–110, **115,** 115–117
   pre-reading/exploration, 107
   Shuffleboard Hero activity, 107–109, **108,** 112–114, **113**
   Thinking Mathematically: Here's the Kicker, 111, 118
kinetic energy. *See* "Energy's Wild Ride"

L

learning cycle, 2–3
"Lighting the Way," 197–210
   application/post-reading/writing, 202
   background, 199
   Code Transmission activities, 199–201, **200,** 203–205
   connections to standards, 198
   journal questions, 202

pre-reading/exploration, 199
"The Sweet Sound of Light" reading, 201–202, **206,** 206–208, **207**
Thinking Mathematically: The Power of Two, 202, 209–210
literacy learning cycle, 4, **5**

M

magnetic fields. *See* "Ding-Dong Electromagnets"

N

Newton's first law of motion. *See* "Inertia: It's a Drag"
Newton's second law of motion. *See* "Kick Force"
new words, 16, **16**
Next Generation Science Standards (NGSS), 2
    lesson connections to
        for "All About Bat Waves," 182
        for "All Charged Up," 78
        for "Ding-Dong Electromagnets," 166
        for "Energy's Wild Ride," 120
        for "Handy Heaters," 60
        for "How to Not Die in Antarctica," 152
        for "Identify This," 44
        for "Inertia: It's a Drag," 92
        for "Kick Force," 106
        for "Lighting the Way," 198
        for "Taking Your Temperature," 136
        for "The Smash-Masters," 24

P

peer conversations, 13
physical properties. *See* "Identify This"
post-reading, 4
    for "All About Bat Waves," 185–186
    for "All Charged Up," 81
    for "Ding-Dong Electromagnets," 169–170
    for "Energy's Wild Ride," 123–124
    for "Handy Heaters," 66–67
    for "How to Not Die in Antarctica," 156
    for "Identify This," 49, 57
    for "Inertia: It's a Drag," 97

Index

    for "Kick Force," 110–111
    for "Lighting the Way," 202
    for "Taking Your Temperature," 140–141
    for "The Smash-Masters," 28, 38–42
potential energy. *See* "Energy's Wild Ride"
pre-reading, 4
    for "All About Bat Waves," 183
    for "All Charged Up," 79
    for "Ding-Dong Electromagnets," 167
    for "Energy's Wild Ride," 121
    for "Handy Heaters," 61–62
    for "How to Not Die in Antarctica," 153
    for "Identify This," 45–46
    for "Inertia: It's a Drag," 93
    for "Kick Force," 107
    for "Lighting the Way," 199
    for "Taking Your Temperature," 137
    for "The Smash-Masters," 25–27
previewing diagrams and illustrations (strategy), 17
    for "Ding-Dong Electromagnets," 168–169
    for "Inertia: It's a Drag," 96–97
prior knowledge, 3

R

reading, 4, 11–12
reading-group roles
    for "The Smash-Masters," 33–34
reading groups, 5–6, **14,** 14–15
    for "The Smash-Masters," 27
reading skill development, 3–4
    approaches to, 12–13
    overarching strategies, 13–15, **14**
    problem-solving strategies, 15–19, **16, 18**
    self-assessment of strategy use, 19–20, **20**
    strategy introduction, 5
reading standards, 2. *See also Common Core State Standards (CCSS)*
reading strategies, 5
reading technical text (strategy), 19
    for "Handy Heaters," 65–66
    for "Kick Force," 109–110
rubrics

    for assessing claims, evidence, and reasoning, **8**
    for evaluating responses to writing prompts, 7

S
safety notes, 8
    "All About Bat Waves," 183
    for "All Charged Up," 79
    for "Ding-Dong Electromagnets," 167, 171, 176, 177
    for "Energy's Wild Ride," 121, 122, 128
    for "Handy Heaters," 61
    for "How to Not Die in Antarctica," 153
    for "Identify This," 45, 46, 50
    for "Inertia: It's a Drag," 93
    for "Kick Force," 107, 108
    for "Taking Your Temperature," 137
    for "The Smash-Masters," 25
science and engineering practices
    for "All About Bat Waves," 182
    for "All Charged Up," 78
    for "Ding-Dong Electromagnets," 166
    for "Energy's Wild Ride," 120
    for "Handy Heaters," 60
    for "How to Not Die in Antarctica," 152
    for "Identify This," 44
    for "Inertia: It's a Drag," 92
    for "Kick Force," 106
    for "Lighting the Way," 198
    for "Taking Your Temperature," 136
    for "The Smash-Masters," 24
science learning cycle, **5**
self-assessment of strategy use, 19–20, **20**
signal words for cause and effect (strategy), **18, 20**
    for "Taking Your Temperature," 139–140, **140**
signal words for compare and contrast (strategy), **18, 20**
    for "How to Not Die in Antarctica," 154–155
standards, 2
static electricity. *See* "All Charged Up"

T
"Taking Your Temperature," 135–150

Index

application/post-reading/writing, 140–141
background, 137
connections to standards, 136–137
"Feverish" reading, 139–140, **140,** 146–148
journal questions, 140
pre-reading/exploration, 137
Thinking Mathematically: Shaking It Up, 141, **149,** 149–150
What Makes Heat Hot? activity, 138–139, 142–145
talk your way through it (strategy), 18–19
for "Energy's Wild Ride," 122–123
for "Lighting the Way," 201–202
teaching notes, 15, 16, 45
temperature. *See* "Taking Your Temperature"
text signals (strategy), 17–18, **18**
"The Smash-Masters," 23–42
application/post-reading/writing, 28, 38–42
background, 24–25
connections to standards, 24
"Fair on the Field—and in the Lab" reading, 26, 29–30
journal questions, 28
pre-reading/exploration, 25–27
Racing for Control: Designing a Controlled Experiment lab sheet, 26–27, 31–32
safety notes, 25
"The Smash-Masters" reading, 27, 35–37
Thinking Mathematically data interpretation, 28, 38–42
think-alouds, 12–13
Thinking Mathematically, 6–7. *See also specific lessons*

W

waves. *See* "All About Bat Waves"; "Lighting the Way"
word definitions, 184, **185**
writing prompt, 7, **7**
writing standards, 2. *See also Common Core State Standards (CCSS)*

www.ingramcontent.com/pod-product-compliance
Ingram Content Group UK Ltd.
Pitfield, Milton Keynes, MK11 3LW, UK
UKHW060443150426
5217IPUK00031B/2098